More evidence - Flying Saucer Review v. 26 #1 p. 29
W9-BFL-104

Life on Mars

Life on

E.P. DUTTON · NEW YORK

Mars

David L. Chandler

Illustrations by Stephen Sluiter and Jane Clark.

All photos not otherwise credited are courtesy of NASA.

Acknowledgment for use of the epigraphs is made to the following: Chapter 2, © 1977 by Fred Warshowsky. Reprinted by permission of Reader's Digest Press · Chapter 3, © 1950, renewal 1977 by Ray Bradbury. Reprinted by permission of Harold Matson Co., Inc. · Chapter 4, © 1976 by Princeton University Press, reprinted by permission · Chapter 5, © 1977 by Robert Jastrow, reprinted by permission of W.W. Norton, Inc. · Chapter 6, © 1956 by I.M. Levitt, reprinted by permission of Holt, Rinehart and Winston · Chapter 7, © 1971 by Jacques Monod, reprinted by permission of Alfred A. Knopf Inc. · Chapter 8, © 1976 by the British Interplanetary Society Limited, reprinted by permission · A portion of Chapter 5, in somewhat different form, appeared in the *Atlantic Monthly,* June, 1977

 · For information contact: E.P. Dutton, 2 Park Avenue, New York, N.Y. 10016 · Library of Congress Cataloging in Publication Data · Chandler, David Luscombe · Life on Mars · 1. Life on other planets. 2. Mars (Planet) 3. Viking Mars Program. I. Title · QB641.C48 1979 574.999'23 78–16656 · ISBN: 0-525-14560-5 · Published simultaneously in Canada by Clarke, Irwin & Company Limited, Toronto and Vancouver · Designed by Nicola Mazzella · 10 9 8 7 6 5 4 3 2 1 · First Edition

For Fay and Brooke,
with love

Contents

Acknowledgments

I WOULD LIKE to express my deep gratitude to some of the people whose help was invaluable in the preparation of this book: Robert Jastrow, for many valuable discussions, and for his early encouragement without which this book might never have come about; Gilbert Levin and Alexander Rich, of the Viking biology team, for their thoughtful answers to my many questions; Paul Solman and Jan Freeman, whose advice and assistance helped to make this book possible; and Don Eyles and Larry Nile, for their careful critical readings of the manuscript.

Life on Mars

In the Beginning . . .

To consider the Earth the only populated world in infinite space is as absurd as to assert that in an entire field sown with millet only one grain will grow.

—Metrodoros the Epicurean
circa 300 B.C.

SCIENCE IS a beacon, casting its illumination into the shadowy recesses surrounding the realms of the known. It enables us to probe deeply into the murky depths of the universe, and to reach back through time to the emergence of the first living things from the primeval ooze, back to the birth of the sun and stars and galaxies, back to the very formation of the universe.

This unraveling of the secrets of our origins, satisfying one of the deepest cravings of the human intellect, did not really get underway until the middle of the last century. Its beginning is perhaps best marked by the publication, in 1859, of Charles Darwin's "The Origin of Species." Until that time, questions about the creation of man and his world had been the exclusive dominion of philosophers and theologians. It was taken for granted that no hard evidence could ever be brought to bear on such cosmic questions.

The theory of evolution profoundly changed the world,

as Darwin knew it would. Suddenly nature was revealed as a flowing process of constant change and growth instead of as a static, immutable order, unchanged since the moment of creation. And the exuberant variety of living things we see around us is surely all the more awesome for being seen as but one stage of an unfolding story whose future is beyond imagination.

Not long after Darwin described the process of evolution by natural selection, others began to fill in the details of the mechanisms that make it work. The chemical components of life have been analyzed in great detail, beginning with Louis Pasteur's discovery of the asymmetry of living molecules and culminating in the discovery of the structure of the DNA molecule, the fundamental unit of heredity, by James Watson and Francis Crick. It may not be long before we are even able to decode the hereditary information in the strands of DNA that make up every living creature on Earth, and thus learn in detail the story of our own evolution.

Evolution is no longer considered by most biologists to be just a theory. Its workings have been so thoroughly documented, and the actual process has been observed so many times under laboratory conditions—albeit on a small scale—that its validity is now virtually unquestioned. The only remaining points of contention concern *how* it works, and such questions are actively under investigation. For example, it is not yet clear whether species evolve at a constant rate, or whether there are sudden spurts of dramatic change.

But whatever the precise dynamics of evolution may prove to be, one dramatic conclusion has emerged to nearly universal acceptance in the scientific community: from the first primitive cells, the laws of evolution led inexorably toward the development of ever more complex, sophisticated, and adaptable forms of life, including ourselves, without any unexplainable leaps that might require a miraculous intervention. The laws of probability are sufficient to account for the whole awesome proliferation of species descending from that primal microbe. Once life began, the emergence of advanced life forms and intelligence was just a matter of time. Different circumstances might have led to the emergence of different particular forms, but the overall process of specialization and diversification are now believed to follow inevitably from the origin of life.

But what of the origin of life itself? At what point was the great divide crossed between nonliving matter and living organisms, and how did that miracle come about? Any answer to this question is necessarily speculative: while the process of evolution can be observed, and sometimes even controlled, no one has yet observed the spontaneous creation of life from inert matter. But, here again, there is a strong consensus among scientists: that life arose spontaneously in the oceans of the newly formed Earth, as a result of natural chemical processes taking place on such a vast scale, over such eons of time, that the improbable finally became inevitable.

In short, most biologists now believe that life is an inevitable consequence, given enough time and the right kind of environment, of the basic physical and chemical laws of the universe.[1]

Once that premise is accepted, it follows that given enough time some kind of life will emerge on any planet capable of sustaining it. In our galaxy alone, estimates of the number of such planets range from the millions up into the billions. One can surmise that the formation of life was probably not unique to the Earth; it may indeed have been commonplace and widespread. We are almost certainly not alone; the Universe may be teeming with life of such diversity that all of the Earth's creatures from artichokes to zebras will seem monotonously alike by comparison.

This assumption follows logically from all that we know about life on this planet. But until now it has been only a theory, supported by a wide variety of indirect evidence, but totally unprovable.

Until now. For the first time in our history, we now have

the ability to explore other worlds directly, in search of evidence that they might be inhabited by something, however strange or primitive, that we can recognize as living. Such a discovery would be exhilarating in itself, and of vast significance: one case of the independent evolution of life outside of the Earth would prove that the process was an easy one. There could then be little doubt that we must be surrounded by a whole host of alien species, some of whom are bound to be as intelligent as we are—or much more so.

The first target in our search for extraterrestrial life was Mars, the most Earthlike planet in the solar system. Mars has been subjected to the most ambitious program of planetary exploration yet undertaken, and so much information has been gathered from the 1976 Viking mission that it will take many years to analyze it fully.

The search for Martian life has barely begun. The Viking landings were just the latest step in a search that is not likely to yield any unambiguous results until the day when astronauts arrive on Mars to begin a direct, systematic exploration—at an optimistic guess, maybe in the 1990s.[2] But already more data has been gathered about Mars than any of our other planetary neighbors, and many important conclusions can be drawn: conditions on Mars are such that life definitely *could* exist there; the Martian climate in the past has been much more favorable to life than it is now, and, most important of all, there is now considerable evidence from the Viking biology tests that Mars is, in fact, inhabited, at least on a microbial level.

These conclusions, and the evidence behind them, will be explored in the following chapters. But in order to arrive at an intelligent assessment of the evidence for life on Mars, we must first answer some fundamental questions about the nature and origins of life itself. How did life begin? How did it evolve to the spectacular diversity that we see today? Are other parts of the universe similar to our own neighborhood, the solar system, so that we may expect to find similar processes taking place elsewhere? What requirements must be met for life to exist on another planet?

To answer these questions, let us begin at the beginning, by reviewing the picture that scientists have gradually pieced together of the formation of the universe and its constituent parts and of the evolutionary stages in the formation of life on this planet.

Since this account spans the fifteen billion years since the universe began, it cannot possibly do justice to the subtle controversies of cosmological and biological theory. This summary does represent the prevailing views of the scientific community at the present time, encompassing all the evidence that is currently available. Some of the details will undoubtedly be revised as new discoveries are made, but the overall picture is well established.

The Birth of the Stars

The universe began with a bang. All of the constituents of the universe had been compressed in one inconceivably hot and dense "soup" of matter and radiation. This amorphous mass had a temperature well over a trillion degrees, and was packed together at a density more than a billion times that of the Earth; under these conditions, matter as we know it could not have existed. Then, suddenly, there was an explosion so vast and cataclysmic, so incomprehensibly devastating, that fifteen billion years later we can still hear the noise it made, as a steady broad-band radio hiss.

Everything in the universe is flying apart like shrapnel from a cosmic bomb. It is this high-speed expansion of the universe, discovered in 1923 by the astronomer Edwin Hubble, that led theorists to believe that it all began with the Big Bang.

Hubble's discovery was based on a painstaking analysis of

the relative distances and speeds of hundreds of galaxies—those great swirling clouds of dust and gas set with billions of blazing stars. (Our sun is just a typical star near the edge of a typical galaxy called the Milky Way.) Billions of galaxies are strewn through the universe, and Hubble found that they are all flying apart at speeds that increase with their distance from us. It is logical to assume that if the galaxies are flying apart, they must at one time have been packed tightly together.

The Big Bang theory was vigorously debated for decades: some speculated that matter was being spontaneously created throughout space so that as everything flies apart the average amount of matter in a given volume of space remains the same. This "steady state" theory, which held that there was no beginning to the universe, has been thoroughly refuted by two kinds of evidence for the Big Bang: the constant hissing static that shows up in radiotelescopes, which, because of its uniformity in all directions, must be a faint echo of that primordial blast, and a glowing warmth that pervades all of space uniformly, raising the temperature of the universe to three degrees above the absolute zero that it would have been if the Big Bang had not warmed things up. These two pieces of confirming evidence have now led to the acceptance of the Big Bang theory by the vast majority of astronomers.[3]

The arguments now center on how it will all end. Some claim that the universe, the debris from that cosmic firecracker, will go on hurtling apart and expanding forever, that space will become emptier and emptier. Others believe that the process will someday reverse itself, that everything will start falling back together and eventually collapse into one mass, perhaps to repeat the whole implosion-explosion cycle endlessly, with a fresh new universe created each time. Einstein subscribed to this latter view, but the question is as yet unresolved.

The matter hurtling out of that primeval fireball was

probably in the form of hydrogen and helium, which are the simplest of all the elements. For millions of years, these two elements were all there was, scattered in a slightly lumpy fashion throughout the universe.

The lumpiness of the distribution of this matter, the reason for which is not known, is of crucial importance. If the gas had been distributed with absolute uniformity, the story would have ended there: the atoms would have continued to fly apart forever, dispersing into the depths of space. Fortunately, that was not the case.[4] Because the gas was unevenly distributed, gravity eventually began to pull the atoms together into denser areas. This process had a snowballing effect: as a cloud got denser, its gravitational attraction increased, allowing it to draw in more and more gas from the surrounding space. These primitive clouds of gas became the galaxies that we see today throughout the visible universe. Within these rapidly rotating galactic clouds, there was still a further lumpiness. These smaller lumps, too, began to collapse under the influence of gravity to form individual stars.

The internal fires of the stars were ignited after gravity collapsed the gas into such a small space that it began to heat up. The more it was compressed, the hotter it got. Finally, when it reached a temperature of ten million degrees, the heat and pressure were so great that nuclear fusion began to take place. Hydrogen atoms were fused together to make atoms of helium. In the process, tremendous amounts of energy were liberated, and it was this explosive energy that eventually balanced the force of gravity so that the further collapse of the cloud was halted. Thus, a new star was born.

It is the nuclear alchemy performed by the stars that creates the very elements from which everything else, including life itself, is made. The fusion of hydrogen into helium continues anywhere from a few million years up to many billions of years, depending on the size of the star. When the hydrogen has been used up, fusion of helium atoms begins, making carbon and oxygen (two of the basic constituents of life). But the transition from hydrogen burning to helium burning is not a smooth one: the outer layers of the star expand violently, making the star millions of times larger, at which point it is called a red giant. When our sun reaches the red giant stage, its outer layers will engulf the Earth. It won't happen for at least five billion years, though, and if we are still around then there will be plenty of time to plan our emigration to another star system.

When the helium fuel is burned completely, further readjustments take place and the progression of fusion reactions continues, making all the elements up to iron in succession. At that point, the fusion stops: iron is too stable to produce the energetic fusion reactions needed to keep the fires burning, so at this point the furnace goes out. Gravity again gains the upper hand, and the star collapses to become a tiny "white dwarf," which gradually cools down to a burned-out cinder.

This sequence of events applies only to small stars—those from about four times the size of the sun on down. Larger stars go through one more stage: because of the much higher gravitational forces involved in the collapse of a large star, some complex interactions take place in its center which release colossal amounts of energy. The result is a violent explosion, called a supernova, which makes the star briefly shine more brightly than its entire galaxy. It is in such explosions that all the elements heavier than iron are created, and by which those elements, as well as those previously formed by the star's nuclear fusion, are scattered through space.

The Andromeda galaxy, a close companion of our own Milky Way galaxy. There are billions of galaxies, each containing billions of stars, and they are all flying apart at tremendous speeds. The detection of this "expansion of the universe" led to the theory of the Big Bang.

Supernovas are rare events, and so are the elements they produce; even now, the universe is still 98 percent hydrogen and helium.[5] But the trace of heavier elements has an important consequence: when new stars begin to form from the interstellar gases, the collapsing protostar will now be surrounded by a cloud of gas greatly enriched by these heavier elements, and this cloud will condense to form planets orbiting around the new star. The first generation of stars, formed from the original clouds of pure hydrogen and helium, was not accompanied by planets. But as the stellar life cycle was repeated, succeeding generations of stars were made from richer material, providing the matter for an array of planets, asteroids, and comets. Thus was the solar system born, four and a half billion years ago. The Earth and everything in it were formed out of material that was made in the explosive death of some ancient star, billions of years ago.

The process of planet formation may now have been observed for the first time. A star has recently been seen in the constellation Cygnus which appears to be at the stage right now where the collapsing assemblage of gas and dust has formed a disk around the newly formed star, which has just begun to shine. At the rate the disk seems to be condensing, within a few thousand years it may have coalesced into separate planetary bodies.[6]

It was once thought that the formation of the planets of the solar system happened as the result of a freak accident, such as a near collision by another star whose close approach to the sun caused matter to be drawn out by the star's gravitational attraction. Theories like this one, which would make planet formation a million-to-one shot, have been refuted by the discovery of planets around some of our neighboring stars.

Planets are much too small to be observed directly over the great distances that separate us from even the closest stars, but the influence of a large planet shows up by creating a wobble in the motion of the star as the planet circles around it. By careful study of the motions of nearby stars, it has been shown that, of the dozen single stars closest to us (double and triple star systems, which are more common than single stars, are considered less likely to have planets because of their complex gravitational fields), almost half have planets at least as large as Jupiter—the largest planet of the solar system. Since smaller planets, like the Earth, would have effects that were much too small to be observed, it is reasonable to assume that these stars may, in fact, have a whole family of planets just as our sun does. The closest of these stars, Barnard's star, is believed to have at least three large planets, adding further support to the theory. And astronomers Helmut Abt and Saul Levy have made a statistical survey from which they conclude that nearly *all* single stars have companions that are too small to be detected.[7]

We may soon have direct photographic evidence for the existence of other planetary systems. NASA is currently working on an orbiting telescope which, because it will not be subject to the turbulence of the atmosphere, will be ten times more powerful than the best telescopes on Earth. It should therefore be capable of detecting some of the planets of nearby stars, thereby ending once and for all any dispute about their existence.[8]

It is a basic axiom of cosmology that the universe is essentially uniform, so that events which occur commonly in one area must also occur throughout space. So if most single stars are accompanied by planets in the vicinity of the sun, this must also be true throughout the galaxies. Since our own galaxy, the Milky Way, contains about 250 billion stars, we would expect to find hundreds of billions of planets in this galaxy alone, whose structure and chemical composition would follow the same general patterns.

Within the solar system, there are two basic types of planets: the terrestrial planets (Mercury, Venus, Earth, Mars, and Pluto) and the Jovian planets (Jupiter, Saturn, Uranus, and Neptune). The Jovian planets, unlike the terrestrials, contain vast amounts of hydrogen and helium. They are the giants of the solar system: Jupiter itself is heavier than the rest of the planets put together. None of these planets are believed to have a solid surface. Beneath their atmospheres of methane and ammonia, there is a bottomless ocean of liquid hydrogen. (There may in fact be a small and totally inaccessible rocky core deep in the center.) They are also extremely cold, ranging in temperature from —150 to —220 degrees Centigrade. These forbiddingly alien worlds are unlikely to harbor any kind of life that would be even remotely familiar to us.

The terrestrial planets, on the other hand, are small rocky worlds with cores of molten iron. When first formed, they were little more than cold heaps of interstellar debris. But soon the combination of gravitational pressure and the decay of radioactive elements in their interiors caused the core of the planets to heat up. Eventually the whole planet melted, causing the lighter elements like silicon and aluminum to float to the top, where they formed a thin crust.

For a long time—in the Earth's case, perhaps as long as a billion years—this crust was very unstable, constantly wracked by explosive eruptions that would make our present volcanoes seem insignificant. At that stage it was a very inhospitable place, bubbling with fire and brimstone, perpetually belching forth molten rock from deep beneath the surface, and spewing out gases like methane and ammonia to form the first evil-smelling (and unbreathable) atmosphere.

This basic pattern was followed by the Earth, Venus, and Mars. On the other hand, Mercury and Pluto, because they were much smaller, did not have enough gravitational attraction to hold onto an atmosphere; the gases simply leaked away into space, making the emergence of life there impossible.

The crucial factor in the Earth's formation that allowed the chemical foundations of life to be laid was the arrival of oceans, condensed from the steam blown out of volcanos. Although these oceans may have initially contained only one-tenth of the water that is there now, the amount would have been amply sufficient. In fact, the smaller volume of water may have helped to promote the highly concentrated solutions from which the first organic molecules were formed.

Organic Chemistry: The Precursor of Life

Every living thing on Earth, from the tiniest bacteria to the great whales, has identical components. On a chemical level, the proteins and nucleic acids that make up a slime mold are virtually indistinguishable from those that make up a fruit fly, a giant sequoia, or a human being.

These proteins and nucleic acids are the building blocks of all known life. Proteins, in turn, are made of amino acids, and nucleic acids are made of nucleotides. Amino acids and nucleotides are thus the sand and cement from which the building blocks of life are themselves made; the formation of these molecules is the first stage in the long process that gives rise to life. These molecules themselves are not in any sense alive, however, any more than a pile of bricks and lumber is a house. Life arises much later, after the necessary materials have been assembled.

How do these basic components, amino acids and nucleotides, come into being? This question was first answered in 1953, when Stanley Miller and Harold Urey conducted a famous experiment to try to duplicate the conditions that prevailed on the Earth soon after its formation, before the emergence of life. They took some of the gases that are believed to have formed the Earth's original atmosphere, and sealed them inside a glass beaker. In this beaker they then repeatedly flashed an electrical spark, to simulate lightning from a primordial thunderstorm. The gases were constantly circulated through another beaker containing water, to simulate the mixture of air and water that takes place in breaking waves.

After a week, the clear water in the beaker had turned a muddy brown. The brown goo that had formed was analyzed, and turned out to be composed of a whole array of organic molecules, including several kinds of amino acids. (The term organic is applied to any molecule containing carbon, except for the simplest compounds such as carbon dioxide. While all life is based on organic molecules, not all organic molecules are produced by living things—as this experiment demonstrated.) Similar experiments have been done since then all over the world, with slight variations in conditions and procedures. These experiments have not only produced a wide variety of amino acids, but also an assortment of nucleotides, and thus have proved conclusively that the formation of these compounds in the primeval oceans of the Earth must have occurred easily and rapidly, giving rise to substantial concentrations of organic materials very soon after the oceans formed.[9]

Additional organic material may literally have rained down from outer space. It has now been determined from radiotelescope observations that even in the clouds of interstellar dust from which stars and planets are formed, there is already a variety of organic molecules present, including many amino acids. These are now believed to arise from chemical interactions within the tenuous clouds in the cold and dark void of deep space. This finding is supported by the discovery of organic molecules, again including amino acids, in many meteorites (those chunks of rock that enter the Earth's atmosphere from space, and which are believed to be unchanged since the formation of the solar system). Meteoritic material enters the Earth's atmosphere mostly in the form of dust from meteors that have vaporized in the atmosphere, and at a rate of 100 tons a day. Thus, organic material from space may also have played a significant part in providing the raw materials from which life was to begin. And these materials must have been available not just on Earth, but on every planet.

It is apparent, then, that planets throughout the universe are fairly throbbing with the raw materials from which life is made, waiting for the right conditions to put them together. The implications this has for our search for extraterrestrial life can perhaps best be seen by analogy. Imagine that, instead of looking for life on other planets, we are looking for radios. We have not yet found any, but we have found that the planets are strewn with a blanket of wires, transistors, and knobs. While this certainly would not prove that there are extraterrestrial radios, it would make their existence a lot less unlikely.

But it is the next step that is crucial. How do these components become assembled? What makes amino acids and nucleotides join together to form living organisms? We know that this next step happened here on Earth. By exploring how it happened, we can assess the chances of its also having happened elsewhere, and decide how likely an event life is, that is, how often the miracle that spawned us has been repeated throughout the galaxy.

The Spark of Life

Within the seething caldron that was the early earth, the stuff of the stars, liberally sprinkled about in the atmosphere and lithosphere, were churned, heated and thrown violently together to give rise to life.

—*Fred Warshofsky*
Doomsday: The Science of Catastrophe

THE EARTH is the only planet on which we know for sure that life began, and we cannot easily generalize from one case. Two cases would establish that life is not a unique occurrence, that it was not caused by some special combination of factors whose repetition elsewhere would be unlikely. And two cases in the same solar system would be strong evidence for the theory that life is inevitable on any hospitable planet. But even from our one example, there are good reasons for believing in the inevitability of life. One such reason comes from the laboratory, another from the fossil record.

In the laboratory, scientists have worked with mixtures of organic chemicals which, as we have just seen, were abundantly available in the primeval seas. By adding a biological catalyst to the mixture, they were able to induce these organic compounds to link up in the form of a primitive kind of DNA. (DNA is the largest and most complex molecule known, and is the basis of all terrestrial life. Its coiled strands contain all of the hereditary information—the genes—that determine the shape and structure of an organism, and even the smallest details of its appearance.) Although this was not a creation of life in the laboratory, it was a recreation of the step in the origin of life that had been the most difficult to explain. If simple kinds of DNA formed in the ancient oceans, then it is easy to reconstruct the whole process whereby the complexity of life as we know it arose by the laws of evolution.

Without the catalyst, the reaction does not take place in the lab. The catalyst is itself a biologically produced molecule, one that would not have been around before life began. Does that mean that nucleic acids wouldn't have formed spontaneously? No, it does not.

To understand why not, we must understand what the catalyst does. It is not an active participant in the chemical reaction, that is, it does not go through chemical changes itself. All it does is to speed up the rate at which certain reactions take place. It does not change the *kind* of activity that will happen, only the amount of *time* that it will take.[1] When the catalyst is not present, the formation of nucleic acids is not observed only because it would take too long to be seen in the laboratory. Reactions that take centuries, or even millennia, could easily account for the origins of life, but no scientist can wait around in the lab long enough to watch them happen.

The fossil record gives a kind of direct evidence for how long the process actually *did* take here on Earth. For about the first billion years, the planet's surface was probably molten. This is demonstrated by the fact that no rocks have ever been found that solidified during that era—the oldest rocks on Earth formed about 3.7 billion years ago. After the surface solidified—and that must have been a slow and gradual process—it must have taken a while for it to stabilize and for the oceans to form. And yet, rocks have been found that were formed 3.4 billion years ago, only 300 million years after solidification, which contain fossils of living cells.

The cells found in these rocks appear to be blue-green algae, a kind of single-celled organism that, like modern plants, breathes in carbon dioxide and exhales oxygen. But experiments by Dr. Carl Woese of the University of Illinois have shown that these were not the earliest forms of life to have existed on Earth. His work, based on "reading" the coded patterns within RNA molecules, has proved that methanogens, organisms that breathe in carbon dioxide and exhale methane, must have come first.[2] In other words, the blue-green algae were themselves the products of a long process of evolution from much more primitive forebears. Since blue-green algae were found in 3.4 billion-year-old rocks, the actual beginning of life on this planet must have happened long before that date.

That means that life must have arisen very quickly (in geological terms) as soon as the planet was ready for it. It could not have taken more than about 100 million years, and may have taken considerably less. This speedy action again argues strongly for the inevitability of the process.

So we now know that at some point about three and a half billion years ago some nonliving molecules first linked together in a form that could be called "life," and thereby began the evolutionary process that has led to the astonishing diversity of living things that we see around us. Let us now examine the stages through which this remarkable progression has occurred.

Stage One: Self-Replication

The beginning of life can be defined as the moment when a molecule capable of reproducing itself comes into existence. As soon as a molecule can start making exact copies of itself, then the whole machinery of evolution—reproduction, mutation, and natural selection—immediately comes into play, creating ever more complex forms. Here on Earth, the trick of self-replication was first mastered by a nucleic acid, a precursor of today's DNA.

The DNA molecule is the most complicated assemblage of atoms known to man, and the information stored in one such molecule, if it could be written out, would fill about 500 volumes of the Encyclopedia Britannica. Yet it is too small to be seen even with the most powerful electron microscope. Although the code in which this amazing molecule stores its information is far from being deciphered yet, we at least know the basic "alphabet" in which it is written: a sequence of four different nucleotides—adenine, cytosine, guanine, and thymine. There are two separate strands of these nucleotides, like strands of pearls on a thread, arranged so that each pearl of one strand touches one pearl of the other. Each of these four nucleotides will only pair up with one of the others (adenine will only pair with thymine, cytosine only with guanine) so that if you know the identity of one "pearl," you can state positively the identity of the one that is touching it on the other strand. In this way, the sequence of nucleotides on one strand absolutely determines the sequence on its paired strand. In order to reproduce, the two strands separate from each other. Since they are surrounded, whether in the primeval soup or within the structure of a cell, by all the necessary nucleotides, these will line up against each strand. The correct mates will line up with each segment, since only one kind of nucleotide will fit in a given slot. They will thus reproduce the whole sequence, so that each strand will have acquired a duplicate of the strand that used to be attached to it. The result of the process will be two complete paired strands, each identical to the original pair.

This process will continue, almost indefinitely, to produce exact copies of the original molecule, or almost exact copies. Every once in a while, there will be a flaw in the reproduction process caused by extreme heat or radiation, so that one or more segments will be copied incorrectly.

Such a random disturbance of the reproductive process is called a mutation, and it will almost always make the resulting "offspring" deficient in comparison to its "parent." But occasionally, just by chance, this mistake will have a beneficial result, something that makes the resulting organism better able to survive in the world. That is the key to the whole process of evolution, for when an advantageous mutation does occur, it will be passed along to all further descendants of the organism in which it occurred.

Furthermore, any mutation that substantially enhances the organism's chances for survival is all the more likely to be passed along since its progenitor will be around longer and thus more able to procreate. This is the process of natural selection, the increased chance for being passed on to future generations of any characteristic, arising accidentally, which is beneficial to the individual.

But how did the very first DNA molecule come to be? It is a very complicated molecule indeed compared to the nucleotides that make it up. It is hard to imagine that such complexity could arise from random events, but apparently that is exactly what happened. Just as a million monkeys at a million typewriters will eventually produce the complete works of Shakespeare, the laws of probability dictate that given enough time the rich organic broth of the Earth's primeval oceans would sooner or later produce a molecule capable of self-replication. The raw materials were available on a vast scale: billions upon billions of organic molecules being formed each *second,* on every square inch of ocean surface,[3] for millions upon millions of years. Also, specialized conditions may have improved the odds in certain places, such as a shallow pond from which most of the water may have evaporated in hot weather, thereby causing the concentration of organic substances—which in the ocean has been calculated to have been about that of a thin consommé—to reach the consistency of a thick, rich soup.[4] And it only needed to happen once: as soon as the process began, that one molecule would keep on making copies of itself as long as the raw materials were available. Life, in its broadest definition, would have begun. But it would be a very crude and inefficient kind of life, unprotected from its surroundings and dependent on a supply of nucleotides in order to continue its self-replication.

The next crucial step that this reproducing molecule must take in order for life to continue and to develop is to separate itself from the outside world, both as a protection against destruction by outside forces, and as a way of maintaining a precisely controlled environment.

Stage Two: The Cell

The cell, or single-celled organism, that we find today has a remarkably complex structure involving many parts working efficiently together. The most important of these parts is a very sophisticated membrane, guarding the border between the cell and its surroundings and selectively allowing only certain kinds of molecules to enter, and certain others to escape. The exact methods the cell wall uses to make these distinctions remains one of the mysteries of biochemistry.

It was the development of this selectively permeable membrane that made further evolution possible, by allowing the chemical activity of the cell to take place in a controlled environment. Because of its complex structure, the cell is able to function as a self-contained factory. The cell membrane allows the necessary raw materials to enter, so that the proteins necessary for the cell's growth and functioning can be made from them. It also lets the waste products from this manufacturing process escape to the outside.

When certain kinds of organic substances are dissolved in water, they immediately form cell-sized globules whose filmy boundaries allow some molecules through, but not others. That may be how cells originated: at some point, a self-reproducing molecule may have found itself inside one of these spontaneously-occurring, nonliving globules. Since this protected environment would have enhanced the molecule's chances for survival, a kind of partnership may have been formed that was to be the precursor of today's cells.

Inside the cell, the main activity is the manufacture of a great variety of proteins. Some of these are enzymes, or catalysts for various chemical processes, such as breaking down ingested food into useful parts and then reassembling them into energy carriers for the cell's later use. Other proteins have structural functions, making up the cell wall, for example, or the tiny hairlike cilia that control the movements of some bacteria.

At this point in their evolution, single-celled organisms were still dependent on a steady supply of organic molecules from the outside in order to continue functioning. The supply of these chemical foods would eventually have been used up. In order to continue to develop and proliferate, the cell had to learn to manufacture its own food. We don't know exactly how this development came about, but it undoubtedly was a gradual process over a long period of time. The result, which made further evolutionary progress possible, was photosynthesis.

Photosynthesis uses the power of sunlight to build complex organic molecules like carbohydrates by breaking down water molecules into hydrogen and oxygen and then combining the hydrogen with carbon dioxide from the air. The oxygen is then released back into the air, and this is the source of all of the oxygen in the Earth's atmosphere, and thus made possible the development of all the air-breathing creatures of this planet.

It is believed that photosynthesis began with the blue-green algae, about three billion years ago. It now is the basic supplier of energy for all living things, directly or indirectly. Every plant cell derives its energy from chloroplasts, small pockets within the cell where photosynthesis takes place. All animal cells, on the other hand, derive their energy from mitochondria, similar pockets in which the products of photosynthesis, ingested by the organism from plant sources, are broken down chemically to provide the fuel the cell needs to function. Thus, the development of photosynthesis was a key that unlocked the door to further evolutionary progress.

After the appearance of the first photosynthetic algae, the rest of the process of evolution is relatively straightforward and well understood. That is not to say that it was easy; in fact, all of the most complicated and interesting developments of terrestrial biology were still to come. But the basic mechanism of evolutionary progress remains the same from the cell on up. There are no more random combinations of chemical components, no more chance encounters of necessary prebiological substances. The element of chance still exists, but only as a small factor in a precisely ordered scheme of reproduction, mutation, and natural selection.

Stage Three: Multicelled Organisms

Once the evolutionary wheels have been set in motion, then the development of all the extraordinary diversity of living creatures that we see on Earth today becomes almost inevitable. It is not that these specific forms are inevitable; the myriad forms of earthly life have been molded by millions of random events, of false starts and blind alleys, and of adaptations to particular conditions at a particular place and time. A few randomly different choices at the early stages of evolution might have produced a world populated by beings totally different from any now existing, a world in which not a single creature would be familiar.

It is not the particular products of evolution that are

inevitable. Rather, it is the whole process and direction of evolution, the progress toward forms that are always more complex, always more sophisticated and better suited to their environment that seems to be inevitably dictated by the very laws of evolution.

The increase in the level of control over the environment, in particular, seems to be a fundamental characteristic. This applies not only to the very advanced kind of control exercised by human beings in manipulating every aspect of the environment to bring it into conformance with our needs and desires. It also applies to the ability of some primitive bacterium in the ancient oceans to move around in a deliberate search for food, instead of drifting at the mercy of the currents and relying on whatever nutrients were swept by. It applies to the gradual development of organs of sight, thereby enabling organisms to exercise greater selection in their movements in search of food, and so on up the evolutionary ladder. Every new sensory apparatus, every new means of locomotion, every improvement in speed of reflexes or in adaptive behavior or in manipulative coordination can be viewed as a way of exercising greater control over the environment, of being slightly less at the mercy of chance events. It is in this sense that control and complexity are the driving forces of evolution.

Some amazingly clever adaptations have occurred that allow life to proliferate and diversify so that it now occupies every available niche on this planet. Species have been able to adapt after millions of years of selection to environments that must at first have been lethal.

At the deepest regions of the ocean floor, for example, a variety of scavenging fish and crustaceans have been found, even though the pressure there is enough to crush the life out of any of the familiar creatures of the sea, and there is no light whatsoever and very little oxygen. Other fish survive in arctic waters whose temperature is close to freezing. There are even some insects that live at temperatures well *below* freezing; they have evolved a kind of organic antifreeze that allows them to flourish when most insects would be frozen solid.

There is a whole class of bacteria that can survive without oxygen; in fact, they are poisoned by oxygen. Other strains of bacteria have been found living and reproducing in the cooling water surrounding the cores of nuclear reactors, in the midst of radioactivity that would be lethal to most living things. And bacteria have also been found breeding in the fuel tanks of jet aircraft.

Blue-green algae are also found under an astonishing range of conditions: in pools of hot sulfuric acid as well as in frigid waters, at high pressure and low. And many single-celled organisms can be frozen indefinitely and then revived.[5]

In short, there is no place on Earth that does not support some kind of living organism. Every extreme that this planet has produced, whether of temperature or pressure or dryness or radiation, has found some creature able to survive it. This has led some people to argue that the Earth represents precisely the range of conditions under which life can exist. This seems a rather nearsighted view; it is surely likelier that once life and the process of natural selection have begun, there are few limits to the circumstances to which it can adapt.

In fact, life seems to thrive under changing conditions. It has been shown experimentally that many bacteria, when subjected to rapidly changing conditions such as an oscillation between extremes of high and low temperature, actually grow and reproduce faster than those that are kept in a stable environment.[6]

It may be that drastic changes in the environment exert a strong selective pressure that may be in no small measure responsible for the evolutionary momentum in the direction of higher forms. In a placid and unchanging world, it is possible that nothing more advanced than a bacterium would ever have arisen.

It was probably the evaporation of ponds of water during

an unusually hot period that caused some organisms, in the desperate struggle to cope with dwindling resources, to adapt to life on dry land. A fish that was able to survive longer than the others in a dried out pond, until the next rain restored its surroundings, would be more likely to produce offspring. A fish whose fins were strong enough to allow it to heave itself overland to the next pond would be even more likely to survive. After a few million years, the transition from fish to amphibian would have been complete.

Similarly, a change to colder conditions may have killed the dinosaurs, while allowing mammals, with their relatively large brains and quick reflexes, to survive.

And it may have been another climatic change that caused the forest lands to dwindle and forced our ancestral apes to venture out into the plains, where the agility and cooperation needed for hunting created a strong selective pressure, and may have been the root cause of the great advance in intelligence that followed.[7]

Stage Four: Intelligence

It was about one million years ago that evolution produced its grandest creation, a thinking animal, the first human. The growth of intelligence was slow and gradual, but we feel an intuitive certainty that there is more than just a matter of degree that separates us from our animal forebears. We are confident that at some point a major threshold was crossed, and that there is an important qualitative difference between us and the lower animals. But just what is this quality that distinguishes us, and that we call intelligence? At what point did we cross the border into humanity?

This is hardly a new question; in one form or another, it has been debated for centuries. Some of the definitions have been merely trivial, like Plato's description of man as the "featherless biped." Other characterizations have provided insight into our nature, but have proved not to be uniquely human. It seems that the more closely we scrutinize it, the fuzzier the borderline becomes.

Man has been defined as the toolmaker, and this has often been cited as our distinguishing characteristic. And, indeed, it seems that the invention of toolmaking was one of the major steps that propelled humanity forward into civilization. Put another way, it is the ability to create one's own environment rather than be dominated by it. As Jacob Bronowski has put it, "Among the multitude of animals which scamper, fly, burrow and swim around us, man is the only one who is not locked into his environment."

The harnessing of fire, of the wheel, the lever, the bow and arrow, of boats and chariots and bridges and viaducts, are a very large part of why we are so different from all the other creatures around us. And certainly the more intricate devices of technology demonstrate a high order of intelligence. But are such artifacts a necessary part of intelligence?

One way of approaching this question, and of gauging the usefulness of our definitions, is to imagine that we have just encountered a race of alien beings on another planet. How can we judge whether or not they are sentient beings, whether we should consider them as beasts or as equals?

Certainly, if we were to find a planet whose inhabitants were building skyscrapers, riding around in airplanes, and watching color television, we would have little trouble concluding that these beings were intelligent. But can we say that the opposite is true? Would the absence of highways and high-voltage lines and high-rise buildings prove the absence of intelligent life?

Surely not. While the absence of tools among dolphins may make their intelligence debatable, it would be unreasonable, on that basis, to exclude the possibility *a priori.* If we were to find that they have a highly developed language (which is widely believed to be true) [8] with a rich oral tradition comparable to the Odyssey and the story of Genesis, how could we declare them to be less intelligent than us, the descendants of toolmaking apes?

Conversely, toolmaking in some primitive forms seems to exist among animals that we do not consider as highly evolved as ourselves. Chimpanzees are known to use sticks to gather fruit that is beyond their reach. The woodpecker, with a brain scarcely larger than a pea, sometimes uses pine needles to reach into holes too deep for its beak, to get at the succulent insects within. Yet a planet of woodpeckers is surely not anyone's idea of the kind of alien intelligence that we might consider our equals.

A more useful index of intelligence is the ability to use language to communicate abstractions. (The qualification is meant to eliminate animal "language" that is nothing more than a set of instinctive signals for attracting mates or sounding a warning.) The invention of an abstract language, even more than toolmaking, has made mankind's progress possible. It is now widely believed that even the rapid evolution of our large brains is directly related to the development of language.[9] Not only is communication impossible without language, but even our very thought processes depend on a language of some kind, whether it be English, Urdu, or algebra. Conscious thought cannot proceed without some kind of abstract symbols to manipulate.

It is hard to imagine a kind of intelligence that could function without language, in its broadest sense. But that is not necessarily to say that we could decipher an alien language, or even that we could recognize very easily whether alien beings used language or not. When we think of language, we think of patterns of sound created by vocal chords and received by ears. Other creatures might achieve the same end by totally different means, some of which we might have a very hard time understanding, or even detecting. Language could be conveyed by any sensory means at all: it could be visual, involving patterns or colors or movements like a kind of semaphore; it could be tactile, like the hand signals used by the deaf; it could even involve tastes or smells, though it's hard to imagine a very efficient communication by such means; or, remembering that we must not look at possible alien life through the blinders of our own experience, it might involve senses that are unknown to us.

Not only might a whole range of senses, known or unknown, be involved in the language of our hypothetical aliens, but they might also communicate through a more direct link —a direct physical connection established by touching foreheads or feelers or antennae or whatever.

In short, while we can expect that any intelligent alien would have a language of some sort, we should not be overly optimistic about our ability to tell whether or not such a language exists, let alone to be able to understand it. The dolphins clearly illustrate this point. There is a great deal of indirect evidence for a highly developed dolphin language, as demonstrated by experiments in which complex sets of instructions, having been taught to one dolphin, were passed along to another by purely auditory links. And yet we have not yet been able to learn a single word of dolphin language, or even to prove conclusively that it contains abstract concepts. The dolphins themselves have done better: some have managed to learn a few dozen words of English, which they can use correctly and consistently,[10] despite the difficulty of using their very different vocal apparatus to make humanlike sounds.

One very objective and empirical criterion for intelligence is the size of the brain, or the number of neural pathways that it contains. The human brain contains about ten times as many neurons as that of our primate ancestors, and about ten million times as many as an earthworm.[11] This seems to be a very clear-cut and precise way to distinguish highly intelligent beings from lower animals. (A large percentage of the neurons in the brains of all animals are used for sensory input and motor control, which require a larger brain size for larger animals. A more precise index of intelligence, therefore, is the ratio of brain size to body size.)

This system agrees closely with our intuition: it classifies humans above other mammals, mammals above birds and reptiles, and birds and reptiles above invertebrates. According to this classification, there is only one other animal on Earth that falls into the same category as man: the dolphin, whose brain is substantially larger than ours, has a brain-size/body-size ratio that makes it exactly our equal.[12]

Communication with this alien intelligence has proved to be a formidable task, despite our similar evolutionary heritage (the dolphin is a mammal), and the fact that they communicate through vocalizations as we do, albeit in a different range. If the day comes when we are faced with an alien intelligence from another world, the difficulties may be greater by a considerable factor. On the other hand, we may hope that someday we will encounter a race of land-based, toolmaking aliens who may surprise us by being easier to communicate with than our aquatic, nontoolmaking cousins.

Life on Other Worlds

Just how likely is it that there is intelligent life—or any life at all—on other worlds? Clearly, any answer to this question is speculative; that does not mean, however, that there are no grounds for making an educated guess. There are ways of narrowing down the uncertainty involved in such speculation, in order to arrive at a reasonable estimate of the odds.

Almost all such estimates are based on a formula first devised by Frank Drake, professor of Astronomy at Cornell University. In essence, the "Drake Equation" [13] is just a way of breaking down the problem into its parts, and assigning estimates of probability to each of those parts. These probabilities are then multiplied together to give the final estimate. Here is a simple version of the formula to determine the number of planets in this galaxy likely to have intelligent life existing on them at this moment: [14]

$$N = S \times P \times H \times L \times I \times T$$

In this formula, *N* is the answer we are looking for, the number of planets that are currently populated by intelligent beings. It is the product of all the variables to the right of the equal sign. *S,* the first variable, is the number of stars in the sample we want to consider. There are 250 billion stars in this galaxy, so we take that as the value of *S. P* represents the fraction of those stars that have planetary systems. The statistical survey of Abt and Levy, mentioned in Chapter I,[15] concluded that this is 20 percent, or one-fifth.

The next factor is *H,* the number of planets in each planetary system that are habitable. In our solar system, we know that two planets fit this requirement: the Earth, and Mars (which has been shown experimentally to be capable of supporting certain algae, bacteria, spores, and lichens).[16] There are other planets that *might* be capable of sustaining life: possible life forms have been proposed for Jupiter, and Saturn's moon Titan is a likely prospect since it has a dense atmosphere and relatively warm temperatures. But these cases are speculative, so we will stick to the two cases that we can be sure of. By the "principle of mediocrity," which holds that our situation is typical of the universe as a whole, we can conclude that two is a reasonable guess for *H,* the average number of habitable planets per system.

L, the next factor in the equation, is the probability that life will, in fact, arise on a planet capable of sustaining it. As we have seen, there are good reasons for believing that this is inevitable, and therefore has a probability of one, meaning certainty. This is in fact the value given to this variable by most scientists who have attempted such an evaluation.[17] We will adopt the value of 1 for now, bearing in mind that this is the most uncertain of all the variables; having only one case to generalize from, we just can't be sure.

The next factor is *I*, the likelihood that life, once it begins, will evolve to a point that we would consider intelligent. Here, we do have more than one case. On this planet, apparently two independent lines of evolution have led to intelligence: the primates (human beings) and the cetaceans (dolphins). We can therefore place greater confidence in our estimate that this, too, is a certainty. Again, we adopt a value of 1 for the variable *I*.

Once intelligence arises, how long will it last? The final factor, *T*, is the length of time that intelligent life continues to exist, on the average, once it has begun. It is expressed as a fraction of the lifespan of the local star. Thus, if man were to survive until the sun explodes, that would give a fraction of a little more than one-half. (The sun, formed about 4.6 billion years ago, is expected to last about 5 billion more.) This is a difficult variable to pin down, but again we can make an educated guess. Man reached his present level of intelligence about one million years ago. (The dolphins have been around much longer. They reached their present intelligence about 30 million years ago.) So a minimum figure would have to be at least one million years. But how long can we reasonably expect mankind to last? It is fashionable these days to adopt the cynical posture that we are close to the end, that we are on the verge of exterminating ourselves through nuclear war, or pollution, or overpopulation. Without minimizing these problems, I think it is likely that man's intellect is capable of extricating him from these difficulties he has devised for himself. But let us examine even the worst imaginable horror: an all-out nuclear war. Civilization might be totally wiped out. Mankind would be dealt a blow from which he might not recover for millennia. But it is utterly inconceivable that every single human being on Earth could be destroyed, and sooner or later civilization—a totally different one, to be sure—would gradually reemerge. Man is not yet capable of utterly destroying himself.[18] So, barring a cataclysm of cosmic proportions, such as the sun blowing up ahead of schedule, it seems that we are here to stay for the next five billion years, unless, of course, we have evolved to even higher forms by then, from which our descendants will look back at us in the same way that we look back upon our progenitors, the apes. So the value of one-half for the variable T seems to be a reasonable one.

What does this all add up to? The formula now contains these numbers:

$$\begin{aligned} N &= 250{,}000{,}000{,}000 \times 1/5 \times 2 \times 1 \times 1 \times \tfrac{1}{2} \\ &= 50{,}000{,}000{,}000 \end{aligned}$$

In other words, based on reasonable estimates from our present level of knowledge, we can expect to find fifty billion planets in our galaxy alone that are inhabited by beings at least as intelligent as we are. That would mean that one out of every five stars has a race of intelligent creatures orbiting around it, so that we could expect to find such a race among the five stars closest to us. This is just an average figure, of course, and there is a great deal of uncertainty in some of the factors, but it clearly shows that the existence of intelligent aliens in nearby regions of space is not nearly as unlikely as most people suppose.

A recently proposed theory makes it seem even more likely. Sir Fred Hoyle, one of the world's preeminent cosmologists, has recently suggested that life may in fact have originated in outer space, in the tenuous films of dust and gas that lie in the spaces between the stars, and from which the stars are born. Hoyle's theory is based on the fact that we know from radiotelescope observations that these interstellar clouds are richly endowed with the organic chemicals from which life formed. He believes that some of these molecules would inevitably accumulate on the surfaces of inorganic dust particles, where they would tend to link together in chains, or polymers.

As the clouds began to condense in the early stages of star formation, collisions between these organic-polymer-coated grains would become more and more frequent. The polymers involved, he has determined, would have approximately the consistency of warm pitch, and so the grains would tend to stick together and form clumps. These clumps would be full of the organic precursors of life.

Hoyle maintains that the whole process that led to the first self-replicating system could easily have happened in these clumps. The same kind of statistics would apply there as in the primeval ocean, considering the vastness of these clouds in space. A Darwinian evolution could have proceeded there, in which ultimately the organic polymer coatings on the clumps of matter might have evolved into biological cell walls.[19]

If this theory seems farfetched, it has received a startling bit of confirming evidence: a study of recently fallen meteorites has revealed that they not only contain a rich assortment of organic molecules, but also an assortment of microscopic cell-like structures that closely resemble the geological microfossils found in ancient rocks on Earth.[20]

So it is possible that life may actually have originated deep in space, and then rained down onto the surfaces of planets throughout the galaxy. This would vastly improve the chances of finding extraterrestrial life, if it is true, because then the difficult first steps in the origin of life would not have had to take place independently on each planet. The seeds of life might have been there waiting to grow on any planet that could possibly accommodate it.

One other consequence of this kind of "panspermia" is that life everywhere would be much more similar, at least on the level of its basic biochemical structures, than if it had evolved separately and independently on each planet. This is also a way to test the theory: if we find alien life, and its chemical constituents are identical to ours, we can be quite sure that we must have had a common ancestor deep among the stars.

But let us return once again to the formula for the likelihood of extraterrestrial life. Of the six factors in that equation, the one with by far the greatest margin of error, the one we can be least sure of, is *L:* the probability that life will arise on any planet that can sustain it.

We adopted a value of 1 for that likelihood, meaning that it is considered a certainty.

But now look back at the previous variable, *H,* the number of habitable planets. We adopted a value of 2, based on the fact that Mars, as well as the Earth, can support certain kinds of life. Taking these two assumptions together, if Mars can support life, and if life is certain to arise on any planet that can sustain life, then it follows that life should have arisen, or must someday arise, on Mars.

What this really means is that we now have at our disposal a clear-cut method of determining the accuracy of the prediction that life is certain to arise if the conditions permit it. If we do in fact discover life on Mars, the theory will have been clearly established. If, on the other hand, we find that Mars is uninhabited, this assumption will have been dealt a serious blow, and we may find that we are much more alone in the universe than we had thought. Thus, the search for life on Mars has implications far beyond that planet.

If we find any kind of life at all on Mars, no matter how simple and primitive and dull it might be, it will make us virtually certain that the universe is teeming with life, swarm-

A particle of ordinary ragweed pollen—the scourge of hay-fever sufferers—seen through a microscope.

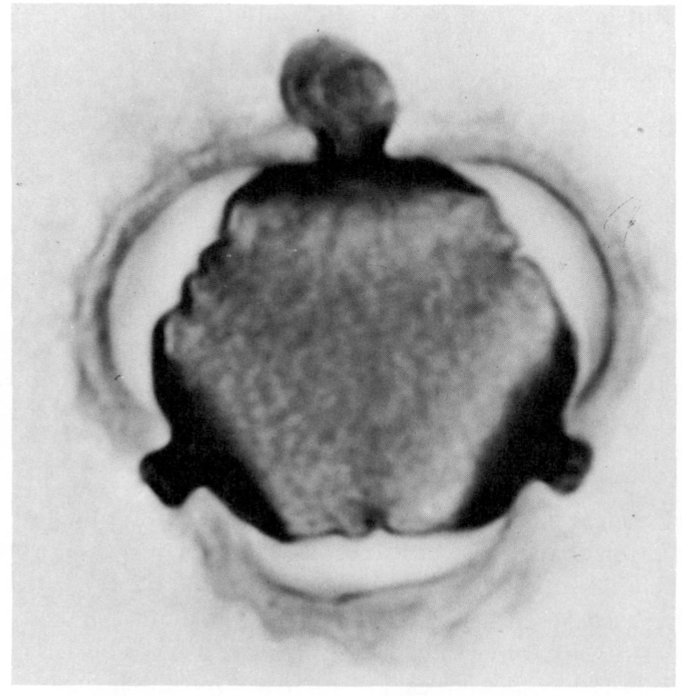

Photos by Edward Anders and Frank Fitch. Reprinted from *Science*, June 7, 1963. By permission of the American Association for the Advancement of Science.

An "organized structure," or microfossil, found inside a meteorite, also seen through a microscope. The astonishing similarity of this structure to the pollen grain (a product of terrestrial life) strongly suggests that there may have been living organisms in the regions where the meteorite formed, in deep space, billions of years ago.

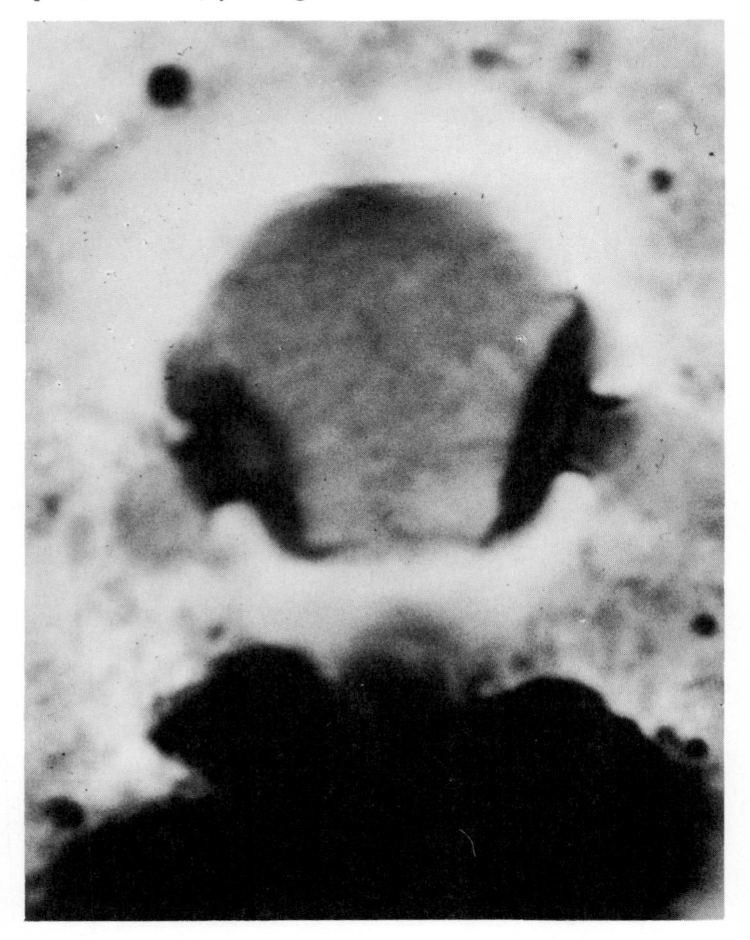

ing with living things of every imaginable kind, and no doubt many unimaginable kinds. We could then be sure that beings equal to us in intelligence abound throughout space, and that beings unimaginably superior to us may also surround our tiny solar system.

Conversely, if there is no life on Mars, we will suspect that in all the vastness of space, man may be a unique creation. We will know that we may be utterly, frighteningly alone.

Either way, it will be a sobering and breathtaking realization.

Mars: The Habitable Planet

"How do you like Mars, Pop?"

"Fine. Always something new. I made up my mind when I came here last year I wouldn't expect nothing, nor ask nothing, nor be surprised at nothing. We've got to forget Earth and how things were. We've got to look at what we're in here, and how different *it is."*

—*Ray Bradbury*
The Martian Chronicles

MARS IS A frozen desert. Its landscape of unrelieved sand and rock resembles the Sahara, but the ground is perpetually frozen like the barren wastes of Antarctica. It is a severe world whose cold, bleak aridity does not suggest a welcome environment for life as we know it.

As harsh as Mars may be, it is by far the most pleasant and Earthlike world in the solar system, and experiments which duplicated Martian conditions on Earth have shown that certain kinds of terrestrial life could probably survive there. Mars is the only other world we know of that is, if not exactly clement, at least potentially habitable.

No other planet even comes close. Mercury and Pluto, like our moon, have no atmosphere at all. The sun shines down on them from an inky black sky, revealing a lifeless expanse of gray rock. Their surface temperatures are highly inimical to life: Pluto averages a frigid −500 degrees Fahrenheit (only 90 degrees above absolute zero), while Mercury goes through daily fluctuations from a nighttime low of −350 to a daytime high of 700 degrees—extremes that no known life form could survive.[1]

The atmosphere of Venus, with its crushing deep-sea-like pressure, creates strange optical illusions, so that its surface looks like a giant bowl with the observer at its center; distant objects appear to be arranged at various points upward on an illusory surrounding cliff that seems to reach right up to the perpetual overcast of steel-gray clouds.[2] An extreme "greenhouse effect," in which the heavy carbon dioxide atmosphere traps the sun's heat much faster than it can be radiated back out into space, creates a sizzling surface temperature of 960 degrees Fahrenheit—hot enough to melt lead. The cloud cover prevents any glimmer of sunlight from ever reaching the surface, and there is a perpetual drizzling rain of sulfuric acid everywhere on the planet. It is surely as close to a vision of Hell as any planet could be.

Jupiter, Saturn, Uranus, and Neptune have no surface at all; above a turbulent shoreless ocean of liquid hydrogen, swirling clouds with vivid, startling colors fill an atmosphere of noxious gases, perhaps broken occasionally by a snowfall of ammonia crystals. The temperatures never rise above −300, and these planets are bathed in a constant field of lethal high-energy radiation.[3]

In this assortment of strange and forbidding worlds, Mars clearly deserves to be called our sister planet. While no planet can be ruled out as a possible habitat for some exotic kind of life, Mars alone could support living things of a familiar kind. True, it is colder than the Earth; but its range of temperatures overlaps ours, and its average temperature is no lower than in the environments some earthly creatures choose to inhabit. True, its atmosphere is much thinner than ours, but it is still thick enough to carry some clouds of water vapor. True, there is no oxygen in the air, but neither are there any poisonous gases. True, the ultraviolet radiation striking the Martian surface is strong enough to kill most living things, but at the time when life began on Earth the radiation here was even stronger. So, while Mars cannot yet be considered prime real estate for earthlings, it is by far the most pleasant place we have yet found beyond the limits of our own warm and fertile world.

It was only natural, then, that Mars should have been chosen by NASA scientists as the target for mankind's first attempt to find life on another world. With this objective in mind, almost a decade of effort and planning were invested in a program which culminated in the launches, on August 20 and September 9, 1975, of the Viking Martian probes.

These twin spacecraft had to cross more than 400 million miles of empty darkness before reaching their objective, a distance more than a thousand times greater than from here to the moon. For comparison, imagine the trip from the Earth to Mars being represented by a flight from New York to Los Angeles. At that scale, the trip to the moon would be a quick cab ride from the East side to the West side of Manhattan.[4] It is no wonder that the Vikings took almost a year to complete their journey, arriving in the summer of 1976.

Each Viking consists of two parts: an orbiting satellite to photograph the Martian surface at high resolution as it whirls around the planet, just as our weather satellites photograph the Earth; and a lander, which separates from the orbiter and then hurtles down to the Martian surface, slowing its descent by firing retro-rockets and deploying large parachutes. The first task of Viking I when it went into orbit around Mars on June 19, 1976, was to take detailed photographs in search of a suitable landing site. The general vicinity for each landing had been chosen before the launch, but within that area the specific site had to be carefully selected to minimize the risk of a crash landing. Three Russian probes had previously been destroyed on impact with the Martian surface,[5] and NASA was understandably anxious not to have its billion-dollar investment end up as a useless scrap-heap of bent metal sitting silently, forever beyond communication, on the desolate surface of Mars.

Since cautiousness was the foremost consideration in the choice of a location for the lander's touchdown, the search was directed toward finding the smoothest, and thus probably the least interesting, of all the possible sites. Toward this end, data were collected from the orbiter cameras and from Earth-based radar scans conducted by the gigantic radiotelescope at Arecibo, Puerto Rico, the largest such instrument on Earth.[6] After rejecting several potential locations as too rough, a smooth area was finally found in an area called Chryse Planitia, or the Plain of Gold. On July 20, at 11:53 AM (Greenwich Mean Time), Viking I finally made a successful landing, precisely on target. Within an hour, the first photograph ever taken from the surface of Mars had been radioed back to Earth, processed, and displayed by the jubilant scientists at Viking headquarters. They had every reason to celebrate; the accuracy of the landing was equivalent to firing a rifle in New York, and having the bullet strike a one-foot target in Los Angeles. Not bad for a first try.

Reckoned by Local Mars Time, the landing had occurred at about 4 PM on the seventh day after the summer solstice. The Martian day is almost identical to ours: 24 hours and 37 minutes. The seasons are also similar to ours—the Martian axis, like Earth's, is tilted about 24 degrees—but they last twice as long since Mars has a year of 687 (Martian) days. So the first six months of the Viking mission were to take place in the Martian summertime.

That first summer's evening on Mars, mankind's robot emissary scanned its environs on this alien world. Standing on its three stubby legs, it began to activate all of its electronic senses. It rotated its two eyes, the twin cameras that can give 3-D views of the landscape, to scan the horizon. With its mass spectrometer, it sniffed the air to determine its composition. It raised an arm into the air to monitor the local weather: temperature, wind speed and direction, and barometric pressure. It immediately began to tell us what it had found by relaying messages, through the orbiter overhead, to the scientists waiting on Earth. Through the information provided by this highly adept robot, the scientists were able vicariously to experience the sensation of actually having landed on Mars, of standing there on its ruddy soil and gazing around at the rocky plain. Their clever machine was able to do most of the things that a human would do to examine its surroundings, with one important exception: it cannot move around. As time went on, this limitation was to become increasingly frustrating.

What it lacks in mobility, the Viking lander makes up for in acuity of vision. The detailed pictures it sent back are so vivid as to make you feel that you have been there, and their value to science is incalculable. Before this landing, the smallest object ever seen on Mars was 100 meters across—about the size of a football field. Viking I's photos, by contrast, were able to reveal objects as small as one millimeter across—100,000 times better resolution, albeit in a very small area, than had previously been achieved. And by now these pictures have been published so extensively that the vicinity of the Viking I lander is probably more universally familiar than most terrestrial landscapes!

These photographs reveal a strange blend of the familiar and the alien. At first glance, they could easily be mistaken for views of some earthly desert landscapes, with reddish sand, boulder-strewn terrain, and occasional dunes. Many people have compared these views to parts of the Southwestern U.S., such as Death Valley or the Painted Desert. The similarities are striking.

The Viking lander. About the size of a compact car, the three-legged lander is so energy-efficient that all of its complex machinery, including cameras, transmitters, sampler 'arm, and more than a dozen sensors and test devices, are powered by a single fifty Watt generator—less than the power used by most household light bulbs.

Photo by Edward Anders and Frank Fitch. Reprinted from *Science,* June 7, 1963. By permission of the American Association for the Advancement of Science.

The view from Viking I. This rocky plain looks so much like some desert landscapes on Earth that it's hard to remember that it is really 200 million miles away.

There are also substantial differences. The most apparent of these is the bright salmon color of the Martian sky. This eerie pink glow gives the appearance of a dawn that lasts all through the day, except that the fringe of pink extends all the way around the horizon. Overhead, except for an occasional bright wispy cloud, the thin atmosphere leaves a dusky blackness that adds to the daylong appearance of a predawn sky.

Because Mars is half again as far from the sun as we are, the sun's disk is only half as large in the Martian sky as in the Earth's. So while its summer is nearly twice as long as ours, it is not as warm because the sun's rays are so much weaker.

At high noon in midsummer at the Martian equator, the highest temperatures may reach 70 degrees Fahrenheit—a nice warm day. But a few hundred miles from the tropics, or a few days from midsummer, or a few hours from noon, the warmth quickly subsides. The average temperature at the

surface of Mars is —15 degrees Fahrenheit, as compared to Earth's average of 75 degrees Fahrenheit. Its extremes are a bit more severe: the lowest temperature detected on Mars, at night in midwinter near the South pole, was 275 degrees below zero, as compared to the lowest temperature ever recorded on Earth, which was 116 below zero in Antarctica.[7]

On Viking I's first day on Mars, the range of temperatures went from an early afternoon high of —23 to a low just before dawn of —123. While this would certainly be uncomfortable for us, some Antarctic algae are known to be able to live in a similar range of temperatures. So, while Mars is a chilly place, it can hardly be considered uninhabitable.

One reason for its low temperatures, apart from the fact that Mars is further from the sun than we are, is that it lacks the heavy blanket of insulation provided by Earth's atmosphere. Mars does have an atmosphere, unlike the airless moon, but it is less than one percent as dense as the Earth's.

The Martian air is about as thin as the Earth's atmosphere twenty miles above sea level.

But as thin as the Martian air is, it behaves in much the same way as ours. It produces strong winds, some of which may reach speeds of 200 miles per hour or more, exceeding the worst terrestrial hurricanes. These strong winds pick up enormous quantities of dust from the surface, creating gigantic dust storms that sometimes spread out to encompass the entire planet, hiding its surface from view for several weeks.

During calmer periods, clouds of water vapor or ice crystals are frequently seen in the Martian sky, most of them resembling the thin streaks of high-altitude cirrus clouds that are seen here on a clear summery day. These clouds are especially common downwind from some of the larger Martian mountains.

One of the biggest surprises from the Viking orbiter photographs was the discovery that Mars also has fog and mist, especially in the early morning. In low-lying areas like some crater basins and valleys, these fogs form regularly every morning, and dissipate an hour or two after sunrise. This discovery shows that there is more water vapor in the Martian atmosphere, and that it goes through more movements and changes of phase, than most scientists had believed—and this, of course, is good news in terms of the prospects of finding living things there; the lack of water had been thought to be the greatest obstacle for the possibility of Martian biology.

Because it is nearer the North pole, there is much more water vapor in the atmosphere around the Viking II lander, which landed on the Plain of Utopia on September 3, 1976, than there is in the vicinity of Viking I. Viking II, which is on the opposite side of the globe from Viking I, is located at 48 degrees North Latitude, roughly equivalent to St. John's, Newfoundland. Viking I is at 22.5 North, which places it in the Martian tropics, roughly equivalent to the latitude of Havana, Cuba.

The temperatures at the Viking II site go through much greater seasonal changes than those near Viking I. While both sites experience maximum summer temperatures of about 30 degrees Fahrenheit, as winter approaches the daily highs become very different: at Chryse, Viking I records maximum temperatures of —10, while at Utopia the winter highs never top —150.

But except for the difference in latitude, the two sites are almost identical. They are both in low-lying plains which fall into the same geological classification, out of the twenty-odd types of terrain that have been analyzed and described by geologists.[8] They both present a very similar face to the cameras, and one might be forgiven for thinking that Mars was a highly uniform planet with very little topographic variety. Nothing could be further from the truth.

This similarity of the sites is not coincidence: remember that the safety constraints in the selection of landing sites dictated that they be in the smoothest, most uniform areas of the planet; additionally, low elevations were preferred in order to get the maximum benefits from the landing parachutes. So it was inevitable that the two locales, though widely separated, are part of the same geological region. They do not come close to representing the enormous variety of surface characteristics that exist on Mars. It is as though two probes sent to explore the Earth had been programmed to land in Siberia and the Yukon; the barren, icy vistas they found could hardly be considered a fair sampling of the surface of this planet. Martian geography is extraordinarily diverse, and we have barely begun our examination of it.

A Planet of Contrasts and Enigmas

Mars is a relatively small planet; it is half the Earth's diameter, but twice that of the moon. Its topography, too, is intermediate between the moon's arid cragginess and the

In this shot of Mars, taken from the approaching Viking I orbiter, the four dark circles in the upper half are the giant volcanoes of the Tharsis region. The topmost is Olympus. The vertical grooves near the center of the picture are part of the Valles Marineris canyon system.

Earth's exuberant mutability. Despite its smaller size, the surface of Mars nearly equals the Earth's land area because of the fact that our planet is mostly covered with water.

Among the great variety of Martian geological formations and surface topography, there are two principal regions: heavily cratered highlands and smooth low lying plains. Surprisingly, these two regions seem to divide the planet very neatly into two equal halves: the Southern hemisphere is mostly cratered highland, and the Northern hemisphere is dominated by the lowland plains. This striking asymmetry of the planet is hard to explain, but it is interesting to note that the Earth is similarly unbalanced: more than three-quarters of its land mass is crowded into the Northern hemisphere, leaving the Southern hemisphere mostly ocean. Geologists have been unable to account for this hemispheric segregation on either planet, but the similarity suggests that some unknown common process may have been at work in both cases.[9]

The asymmetry of the Martian terrain was responsible for some great surprises and controversies as our exploration of Mars has progressed: when the first successful Mars probes, Mariners 4 and 7 (which were similar to the Viking orbiters, but somewhat smaller), photographed parts of the planet during a quick flyby, by sheer coincidence nearly all of their pictures showed the Southern hemisphere, with its ancient and heavily cratered plains showing little sign of geological activity: erosion, sedimentation, or volcanism, the signs of an active planet. These regions seem to have been preserved with very little change from the earliest stages of planetary formation.

From these pictures, which at the time were by far the most detailed views of Mars ever seen, the disappointed scientists reluctantly began to conclude that Mars was a dead world, not only incapable of supporting any imaginable kind of life, but devoid even of any geological interest. But the Mariner 9 probe, which went into orbit around Mars in order to conduct a detailed mapping of the entire surface, changed all that.

The Mariner 9 mission had an inauspicious beginning, but ended up as a resounding success. When it first arrived in orbit around Mars, in 1971, the entire surface of the planet was obscured by a dust storm, which continued for several months. For this entire period, Mars presented a totally featureless globe to the cameras, and scientists were beginning to worry that all their efforts might have been wasted. Then, gradually, the dust began to settle, and slowly the surface began to be revealed. Eventually, the entire Martian surface was photographed with crystal clarity.

The first features to appear through the murky atmosphere as the dust storm subsided were the highest mountain peaks on Mars: the great volcanos in the region called Tharsis. The largest of these is called Mount Olympus, after the legendary home of the gods of ancient Greece. Amazingly, Olympus was first seen on Mars, and given its name, almost a century ago by astronomers who observed it as a bright white patch (probably as a result of clouds or frost forming around the mountain's peak) and so called it "Nix Olympica," or Snows of Olympus.

Olympus dwarfs every other mountain known in the solar system. The largest mountain on Earth, if measured from the ocean floor to the peak, is the volcano Mauna Kea in Hawaii, which stands 6⅓ miles high; Olympus would tower over it, rising fifteen miles above the surrounding plains. Its base covers an area larger than the state of Washington, and is topped by a caldera, the crater from which bubbling lava once spewed out across the plains, that is forty-five miles across.

The discovery of volcanos on Mars was an extremely important event, because it has great implications for the habitability of the planet. For one thing, it is now generally accepted that the Earth's original atmosphere originated from

Olympus Mons, the largest known volcano in the solar system, is three times higher than Mount Everest.

the gases belched out of the depths of the planet in thousands of volcanic eruptions. And without that original atmosphere, life would never have had a chance to evolve here. Thus, the existence of Martian volcanos indicates that it is at least possible that Mars once had enough of an atmosphere for life to have begun there.

Another important consequence of the discovery of Martian volcanos is that it proves that Mars, like the Earth, has a hot, molten core. This means that it is possible that pockets of heated water, like Earth's "hot springs," may have formed below the Martian surface. If so, they would form an ideal environment for Martian creatures to survive in, avoiding the frigid temperatures of the surface.

Before the Mariner 9 photos, many scientists had doubted that Mars had any volcanos, but now the question has been settled. Now, scientists are trying to figure out why the volcanos there are so much bigger than any on Earth.

One theory is that Mars does not have a surface that is broken up like the Earth's. On this planet, the surface is broken up into dozens of rigid plates, all of which are in constant motion relative to each other. Therefore, wherever there is a volcanically active area in the mantle beneath the Earth's crust, a succession of volcanos is formed as the crust moves over that area (over a period of millions of years). The Mauna Kea volcano is a good example: following a straight line north and west from Mauna Kea, there is a whole string of extinct volcanic cones running for more than 1,500 miles.[10]

So, the theory goes, perhaps there is no motion of crustal plates on Mars. Thus, the volcanic vent of Olympus, instead of creating dozens of smaller volcanic mountains, has piled all its lava up into one gigantic stationary cone. The seismometers on the Viking landers, which are designed to detect Marsquakes, may provide enough information about the internal structure of Mars to either prove or refute this theory.

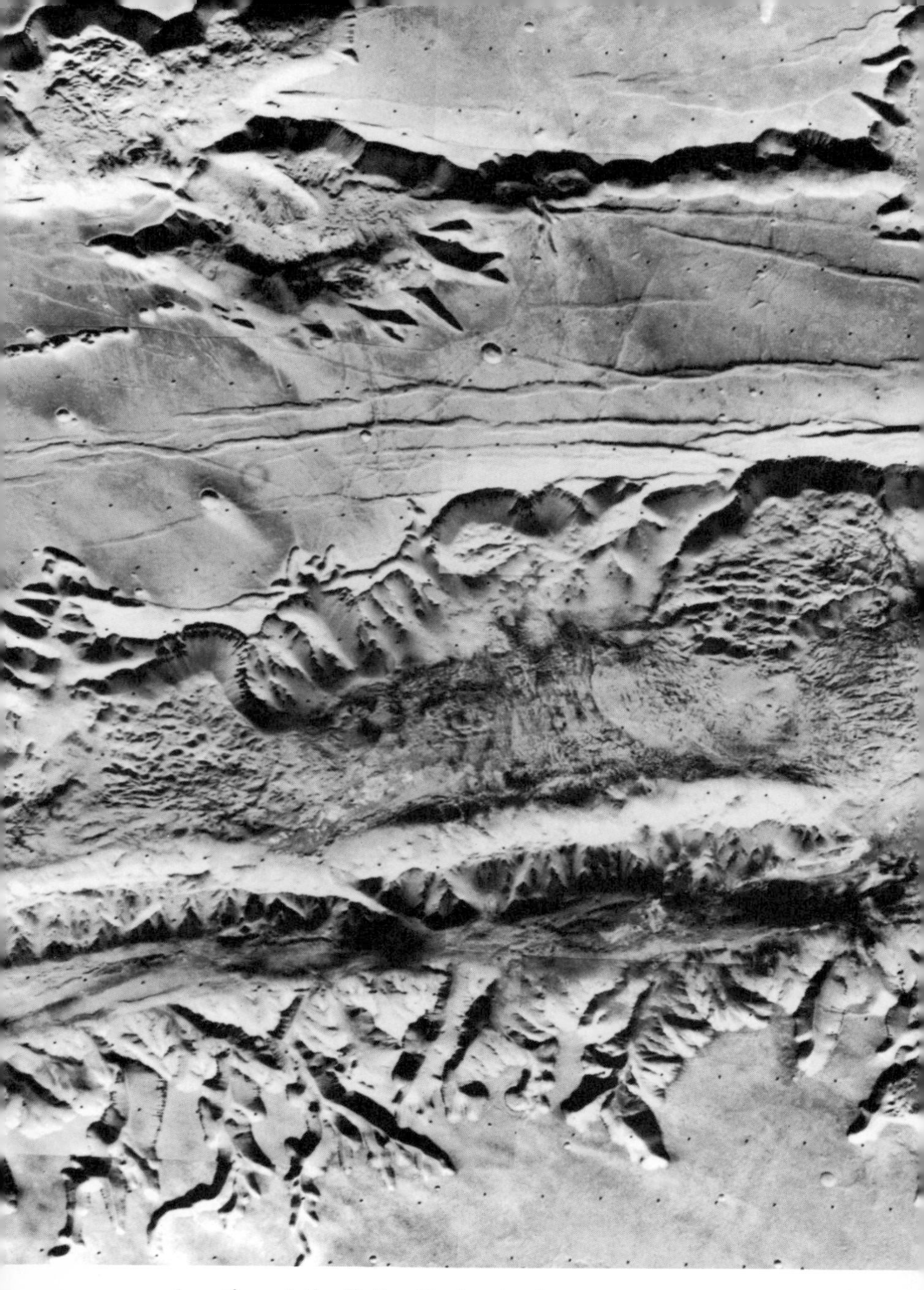

A section of the Valles Marineris, the great canyon system near the Martian equator, which spans a distance greater than the width of the United States and has a depth of four miles.

Another sign of large-scale geological activity near the region of the Tharsis volcanos is Valles Marineris (the Valleys of the Mariners, named for Mariner 9 which first discovered them). It is the largest canyon system ever seen anywhere in the solar system.

This great set of gashes in Mars' side runs for almost 3,000 miles, more than the full width of the continental United States. It connects to a branching system of several lesser canyons which, although dwarfed by Valles Marineris, are themselves larger than any terrestrial counterpart. By comparison, the Grand Canyon seems like an insignificant ditch: parts of Valles Marineris are four times deeper and more than eight times as wide. Massachusetts and Connecticut together could fit comfortably between their sheer cliffs, which stand 150 miles apart at the widest point. Despite their much greater size, the Valles Marineris do resemble the Grand Canyon in appearance: they are bordered by steep cliffs that seem to show signs of having been carved out of layered material, revealing many sedimentary layers along their faces.

The origin of this gigantic rift system is another unsolved mystery. One possibility is that it shows the beginning of the breakup of the Martian surface into tectonic plates. Perhaps the process of crustal fracturing began, but never really got very far because Mars may, as a result of its smaller diameter, have a much thicker crust than the Earth. The thickness of the crust might also explain the great length and depth of the canyons. This hypothesis is supported by the close similarity in appearance of this system to the Great Rift Valley of East Africa, which is known to be the result of a boundary between two of Earth's crustal plates.

If Valles Marineris does represent the beginning of the breakup of the Martian crust, that still leaves unanswered the *reason* for its breakup. Even on Earth, where the existence of many separate surface plates is well established, geologists have been hard put to explain the cause of this fracturing of

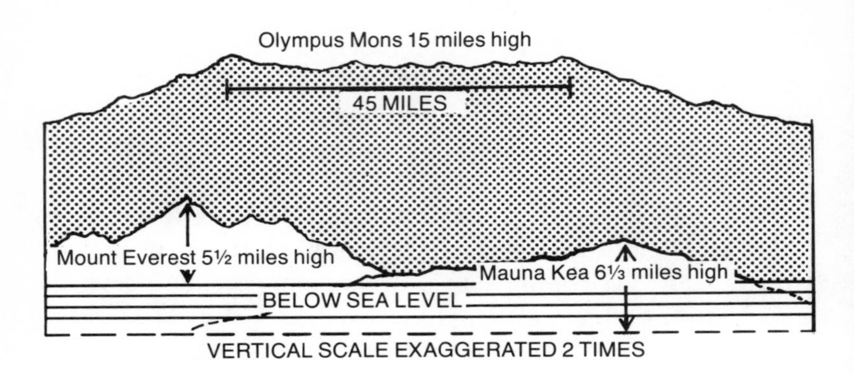

Olympus Mons, the largest of Mars' volcanoes, shown to scale with Mount Everest, Earth's highest point above sea level, and Mauna Kea which, measured from the sea floor, is the tallest mountain on Earth.

the surface which makes it resemble the shell of a hard-boiled egg that is ready to be peeled. One theory is that the original fracture resulted from tidal stresses in the Earth's crust caused by the influence of the moon. If so, that might explain why Mars has experienced much less fracturing than the Earth: Mars does have two moons, but they are much smaller than ours and so have very little tidal effect on their parent planet.

The two moons, Phobos and Deimos, were among the most enigmatic objects in the solar system until they were photographed by Mariner 9. Because they are so tiny, until that time they had only been seen as pinpoints of light even through the world's largest telescopes, so nothing at all had been known about their size, shape, or composition.

Phobos, the larger and nearer of the two (whose Greek name means "fear" and is the root of the word phobia), would appear to a Martian observer as less than a third the diameter of our moon, and less than one-twentieth as bright.[11] Its slightly irregular ellipsoidal shape gives it the appearance of a lumpy potato. Phobos has an orbit so close to Mars that an

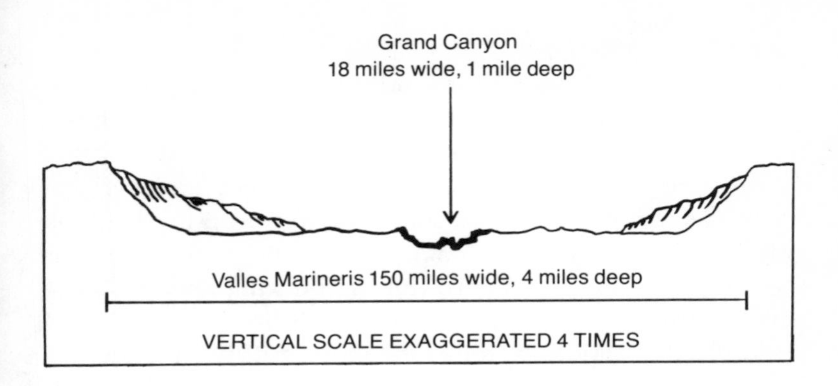

Valles Marineris, the great Martian chasm, shown to scale with the Grand Canyon.

observer within 1,000 miles of the Martian poles would never see it at all; it would always be below the horizon. Compared to our moon, it races across the sky: from its rising point in the west, it moves across the sky and sets in the east just five and one-half hours later. This reversal of normal rising and setting points happens because Phobos orbits Mars faster than Mars rotates. Even though its orbit is in the same direction as our moon's, its rapid motion makes it overtake a given point on the surface so that it appears to go the wrong way.

Deimos (meaning terror) is even smaller than Phobos, and looks like a very bright star, about as bright as Venus looks to us. This unspectacular moon rises in the east and sets in the west, having a rotation period slightly longer than the Martian day. Because its orbit is larger than that of Phobos, it is visible nearer the poles.

Both of these moons are strange objects, unlike the moons of any other planets. They are the smallest and most irregularly shaped moons in the solar system, and bear much less resemblance to the moons of other planets than to the asteroids, those fragments of rock that circle the sun between the orbits of Mars and Jupiter and that were once believed to be the remains of a planet that had exploded, but which are now thought to be remnants from the early, formative stages of the solar system. It may be that Phobos and Deimos started out as asteroids, and somehow were captured by the gravitational field of Mars. Many astronomers think this was the case, but they have a hard time accounting for the almost perfectly circular orbits of both moons: if they were captured asteroids, they should be following narrow elliptical paths.[12]

One earlier theory that had captured the imagination of many has been knocked down by the latest Viking photographs of these two moons. In the early 1960s, the Russian physicist I. S. Shklovskii noticed that Phobos seems to be speeding up in its orbit, indicating that it is getting closer to Mars. His calculations of the expected speedup caused by friction with the Martian atmosphere showed that this was not nearly enough to explain the observed effect, unless the moon were much lighter than it should have been. Since its size was known with a fair degree of accuracy, it seemed that the only way it could possibly be light enough to explain the acceleration was if it were hollow, and the only way it could be hollow was if it had been manufactured as an artificial satellite.[13] For a while, this seemed to be evidence for intelligent life on Mars. But the closeup photographs taken by Mariner 9 showed clearly that Phobos is a solid chunk of rock, and revealed that the original measurements of its acceleration had been in error.[14]

But although the Mariner 9 photographs showed that Phobos was not hollow, thereby ending speculations about possible artificial origins, the Viking orbiter photographs have raised a whole new set of questions: most of the surface of Phobos turns out to be crossed by hundreds of straight, parallel grooves which no one has been able to explain. What could have caused them? One theory is that Phobos is a chunk that broke off from some larger body, and that the grooves are exposed layers of sedimentary material from that body. Another possible explanation is that the grooves are fracture lines caused by tidal stresses in the moon's structure. But we probably will not know for sure what the explanation is until a future Mars mission allows us a closer look.

Another enigmatic feature of Mars that has emerged from the Viking orbiter pictures is the layered structure of the North polar ice cap. This layering gives the appearance of a tall stack of plates, each one slightly larger than the one above it. Each plate seems to be about 100 feet thick, and there are scores of them, stacked up to a great height. Apparently, these layers must be the result of hundreds of shifts in the Martian climate that caused the ice of the polar cap to melt during warmer epochs, and then to freeze again at the pole as the climate cooled down once more. Each time, a certain amount of Martian dust was mixed in with the ice as it froze, and each time the ice evaporated away again a layer of dust was left behind.

The composition of these polar ice caps was the subject of vigorous debate until the Viking I orbiter settled the question. These vast sheets of ice had been thought by many observers to be composed of frozen carbon dioxide, or "dry ice." But detailed temperature measurements made by the infrared sensors on the Viking orbiter have ended the dispute once and for all: the ice is much too warm to be dry ice, which has a much lower melting point than water ice. It's impossible to compute exactly how much, since we do not yet know how thick the layer of ice is, but there must be vast amounts of water ice. Conservative estimates lead to the conclusion that if it were all melted, it would produce enough water to cover the entire planet to a depth of at least a few fathoms.

The supposed absence of water has long been the strongest argument against the prospect of finding life on Mars. Many scientists had believed that Mars was almost totally devoid of water, which would have meant that it must be as lifeless as the moon. But the Viking analysis of the ice caps, by revealing water in greater abundance than even the most optimistic scientists had estimated, has vastly improved the odds for life.

Another major reservoir of Martian water is believed to be in the soil itself, in the form of permafrost like that of the Alaskan tundra. This probably accounts for as much water as there is in the ice caps, and perhaps even more.

Recent observations conducted by Canada's Algonquin radiotelescope have given strong support to this theory. Radar measurements of temperatures beneath the Martian surface were made by beaming radio waves toward Mars from the giant parabolic dish of the telescope. These waves penetrated several inches into the soil before being bounced back to Earth where the telescope recorded the strength of the echo. These measurements showed a large discrepancy in the temperatures of some areas of the Martian surface, which did not seem to be related to elevation, surface texture, or any other identifiable feature. The only explanation the astronomers have been able to come up with is that the discrepancy must be caused by large concentrations of subsurface water in those areas.[15]

While water ice may be abundant on Mars, liquid water is unknown there (at least on the surface). Because of the low atmospheric pressure, water on the Martian surface goes directly from a solid state (ice) to a gaseous state (steam) without ever passing through a liquid phase, just as "dry ice" (frozen carbon dioxide) does on Earth.

Actually, there *is* one place on Mars where water could

This closeup of Phobos, one of Mars' moons, was taken by the Viking I orbiter. It revealed a mysterious set of grooves in the surface whose origin has not yet been explained.

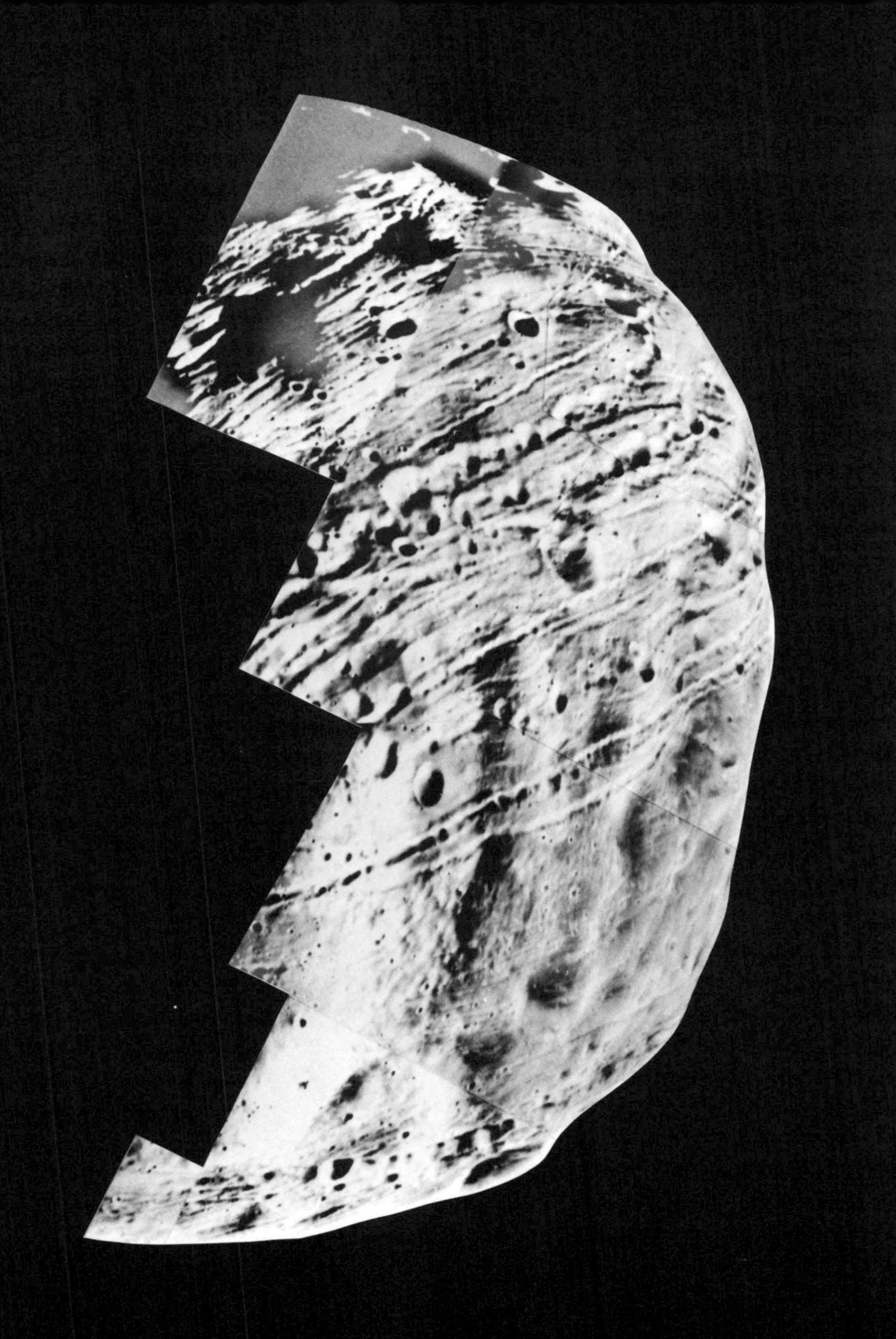

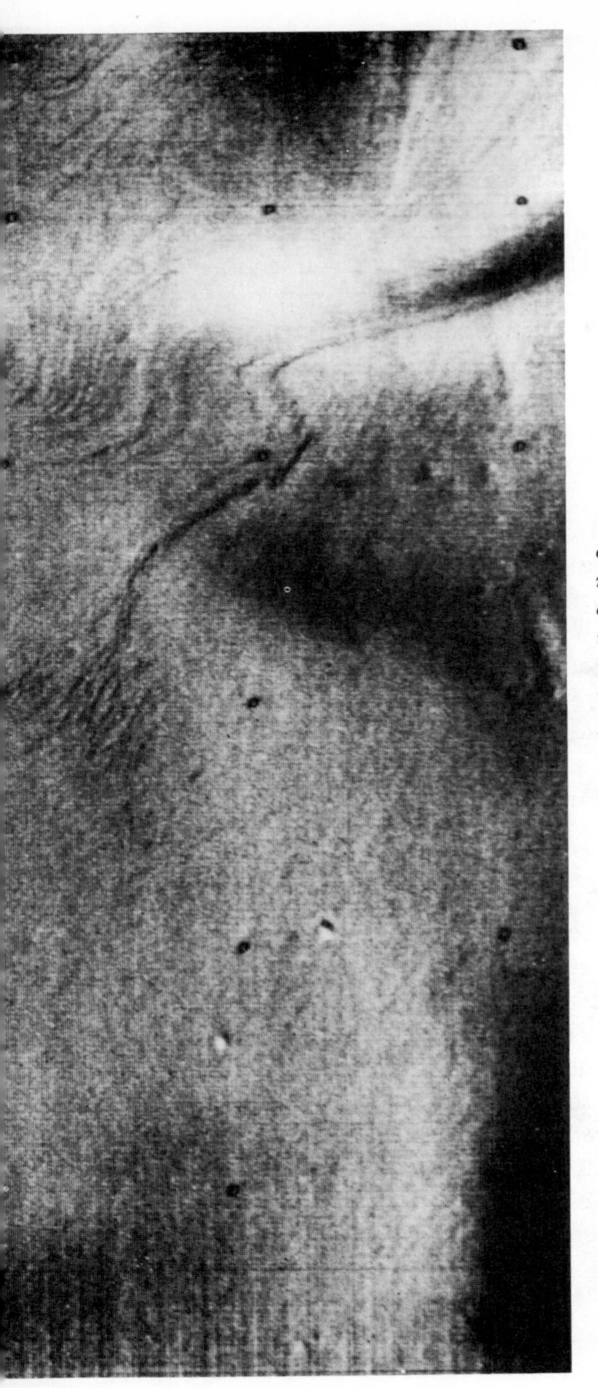

The layered structure of the ice cap at Mars' north pole is another enigmatic feature. It may be the result of millions of years of periodic thawing and refreezing, perhaps similar to the cycle of ice ages on Earth.

exist as a liquid for short periods, if the temperature were high enough. It is the great circular basin near the South pole called Hellas, which is more than a thousand miles in diameter. The deepest part of this basin, which is the lowest point on the whole planet, has a high enough atmospheric pressure for water to remain stable. While no evidence of water has ever been observed there, small local pools may form whenever the temperature rises above the freezing point, as it may during some days in the local summer.

Although there is no liquid water on Mars today, the whole surface of the planet testifies to the fact that it was not always so. Throughout much of the planet there are long meandering valleys, with well developed systems of tributaries. The tributaries all run downhill, as one would expect from river channels. Geologists have been unable to advance any other plausible explanation and have concluded that they must indeed have been rivers at one time. This means that, sometime in the past, the Martian climate must have been very different: there must have been much milder temperatures, or all that water would have frozen; and the atmosphere must have been much denser, or the water would simply have evaporated away in the higher temperatures.

"Mars Jars" and the Prospects for Life

None of the present conditions on Mars in any way precludes the existence of life there. Every hostile aspect of the Martian environment can be met by suitable adaptations that have occurred right here on Earth, even though there has been no great selective pressure in favor of such adaptations. There are some forms of earthly organisms, even though they evolved under totally different circumstances, that could easily survive on Mars. If life ever got a start on Mars, there is little doubt that it could have survived and adapted to the present conditions.

The absence of liquid water on the planet's surface in the present era may be an obstacle, but certainly not a barrier, to the existence of Martian organisms. Water is a fundamental requirement for any kind of life we know of or can easily imagine. But even though Mars is a dry world, there are a number of ways that an organism on Mars could obtain enough water to live.

First, there is some water vapor in the air which could easily be used to satisfy the needs of a well-adapted organism whose requirements for water were minimal. Even on Earth, where water is abundant, there are some species that have become amazingly adept at preserving the precious amounts of water they have.

Many kinds of desert creatures such as the kangaroo rat are able to survive with no water at all except that which they get from their food. And many terrestrial plants are able to store water very efficiently in order to get the most use from a scarce resource. On Mars, where water is at a premium, it is easy to imagine that such abilities could have evolved to a much higher degree in order to cope with the extreme aridity of the place.

Furthermore, there are local and seasonal concentrations where the amount of water vapor in the air is much greater than average, for example around the fringes of the polar ice during the spring thaws. This relatively water-rich environment might be an ideal place for Martian life to thrive.

Second, there is abundant ice on Mars both in the ground and at the poles, so any creature that could melt small amounts of it by using its metabolic heat would be amply supplied with water. Such organisms, although they do not exist on Earth, could easily have evolved on Mars; NASA biologists refer to them as "crystophages," or ice eaters.

Third, there is a great deal of water chemically bound in the surface materials of Mars; these hydrated minerals, in which water molecules are loosely attached to the molecules of the mineral, might give up their weakly bound water in response to heat or a chemical catalyst provided by suitably evolved organisms, which would then be called "lithophages," or rock eaters.

Another obstacle that any Martian organism would be faced with is the intense ultraviolet radiation that reaches the surface throughout the day. Ultraviolet light, or UV, is that familiar "black light" which is used in many hospitals and dentists' offices as a sterilizing agent, because of its lethal effect on microbes. On Earth, these deadly rays are filtered out of the sunlight by the ozone layer, a thin blanket high above the stratosphere that is opaque to UV, but lets the rest of the sun's rays pass freely. Since Mars has no ozone layer, UV reaches the surface unimpeded, at a strength that would be lethal, given a long enough exposure, to any terrestrial life form.

But it is not hard to imagine ways that Martian organisms might have devised to cope with this problem. For example, a large creature could be protected simply by growing a thick opaque shell. Smaller organisms, like microbes, could stay safe from the lethal rays simply by hiding under a grain of sand, or by living inside the spaces of a porous rock as some algae have been found to do in Antarctica. Some organisms may even be able to protect themselves by producing a layer of strong pigmentation to filter out the dangerous part of the spectrum. So, while UV would be a problem for any Earth creature that suddenly found itself transported to Mars, there is every reason to think that Martian organisms, if they exist, would have found ways to evolve and adapt so that this would not be a problem. In fact, the UV may even have had beneficial effects: many biologists believe that ultraviolet rays striking the Earth's surface in primeval times may have been the very force that caused the chemical reactions from which life began. If other circumstances on Mars were ever suitable for it, the UV may have caused life to begin there in the same way.

Some people have considered the absence of oxygen in the Martian atmosphere to be a problem for advanced forms of life. But there are many microorganisms on Earth that survive without oxygen, and which in fact are poisoned by it. They survive in hot springs or in deep mud where there is little or no oxygen to bother them. They breathe carbon dioxide, which is the main constituent of the Martian atmosphere. Such microbes are now believed to be the most ancient form of life on this planet, so we can assume that if there is any life at all on Mars, it would include these "anaerobic" forms.

But even creatures that do require oxygen may have been able to evolve, and survive, on Mars. There is plenty of oxygen available there in the form of carbon dioxide (CO_2) and water (H_2O). On Earth, plants use sunlight to break down water molecules into hydrogen and oxygen. They then release the oxygen into the atmosphere, where animals breathe it in. If there were extensive plant life on Mars, we might expect to find abundant oxygen in the atmosphere there; the fact that there is none has been considered evidence against abundant Martian vegetation. But it does not have to be that way: a symbiotic relationship could easily have evolved in which a plantlike organism coexists with an animal-like one, so that the oxygen released by the plant is transferred directly to the animal without any being lost to the air. Such an arrangement would be a very efficient one for both partners.

The final obstacle for Martian life is the frigid climate. The Martian surface temperatures rarely rise above the freezing point of water, and few terrestrial organisms can survive being frozen. Still, some *can* survive it, and there are some species that do very well at temperatures as low as those encountered on Mars.

The daily range of temperatures in the middle latitudes of Mars is similar to the yearly range in Antarctica, and a wide variety of creatures survive there, including penguins, many kinds of fish, and some simple plants. In the Arctic regions, an even wider variety of creatures have adapted to similar low temperatures, including polar bears and walruses. Even in the tundra, where the ground is perpetually frozen solid just as it is on Mars, layers of moss and lichens are found growing on the ice.

Some Arctic species, including some insects and crustaceans, have managed to evolve a kind of organic antifreeze that protects their body fluids at temperatures well below the freezing point of water. Similar adaptations could enable some organisms to survive on Mars.

In short, there is no theoretical barrier to the possible existence of creatures that could survive the rigors of a Martian existence. But Martian creatures, if there are any, may have an even easier time of it. Even though liquid water cannot exist on the Martian surface, there may be pools of liquid water on Mars just *below* the surface. There could even be underground lakes and rivers, perhaps a whole network of them, caused by geothermal heating of the planet's crust. Many astronomers are convinced that there may be extensive underground water on Mars, and if this is true then such environments would not suffer from any of the obstacles to life that have just been described.

Such subsurface pools would arise when hot magma from deep below the crust rises up through faults in the rocky surface material of the planet. In the process, the surrounding rock is heated up, thereby melting any pockets of ice that were frozen in the ground. Thus, not only could there be liquid water just below the Martian surface, but it could even be *warm* water, which would be a delightfully comfortable habitat that could accommodate any number of familiar creatures, not only algae and bacteria, but also a wide variety of aquatic plant and animal life.

If such a subterranean (or, more accurately, subMartian) reservoir ever broke through to the surface, it might freeze over, forming a thin crust of ice over the top of the pool. In this state, a large pool could remain stable for a long time. This would allow some sunlight to filter through the ice, which would enable plants to conduct photosynthesis and thus provide nutrients for a highly developed food chain. And the ice would, at the same time, screen out all the UV. Such oases in the Martian desert might be teeming with life, even if the rest of the planet's surface turned out to be totally sterile.

But that does not mean that we should give up on the possibility of life under the more rigorous surface conditions; on the contrary, there is every reason to be optimistic. In fact, experiments have shown that some terrestrial organisms are capable of surviving, just as they are, under the very worst of Martian conditions.

The habitability of Mars was first demonstrated in the 1960s, when experimenters duplicated Martian conditions inside sealed glass containers, which of course were called "Mars jars." Inside these jars, they placed a soil composed largely of limonite, the mineral believed to be responsible for Mars' red color. They introduced an atmosphere of carbon dioxide and nitrogen, the principal constituents of Martian air. They shone ultraviolet lights on the jars to simulate the effects of Martian sunlight. They included only the scant amount of water known to be present in the Martian air. They used vacuum pumps to reduce the air pressure inside the jars to the highly rarefied state of the Martian atmosphere. And they chilled the jars in a daily cycle that reproduced the typical range of Martian surface temperatures.

When they introduced ordinary earthly bacteria into the Mars jars, most of them died before the end of the first day's cold cycle. But a significant number survived, and those that made it through the first day were able to survive indefinitely under those hostile conditions. A kind of natural selection had taken place on a small scale. The survivors even began to reproduce.[16]

If terrestrial organisms were able to adapt so quickly to these totally alien conditions, it is certainly not unlikely that Martian organisms, over thousands of generations, might have evolved to a point at which they are perfectly adapted to circumstances that seem nearly lethal to us. In fact, we might expect that, if they exist at all, they are as perfectly attuned to the Martian environment as we are to the Earth's.

Even more advanced life forms have been shown to be capable of surviving some, if not all, of the Martian conditions. Mars jars that simulated some of the Martian parameters, such as temperature and pressure, have been tried on a number of terrestrial organisms, including an ant, a worm, and a larval crayfish; all of them survived. A number of plants have also passed the test, including such unexpected varieties as cucumbers, rye, and millet. One interesting discovery from these simulations was that plants are much better able to survive the extremes of Martian temperature when there is no oxygen at all in the chamber, as on Mars, than when earthly oxygen levels are included.

One of the hardiest kinds of plant life on Earth, and the one considered most likely to be able to survive on Mars, is the lichen. Lichens are found in almost every terrestrial environment: they are the last things seen by mountain climbers as they approach the highest peaks, and they can be found in both desert and arctic climates. In the laboratory, they have been subjected to a total vacuum for a period of years, frozen to 300 degrees below zero, and then revived as though nothing had happened. They are extremely well adapted to low temperatures and dry environments, which makes them ideally suited to Martian conditions. And lichens from perpetually sunny areas have developed pigments to protect them from the solar ultraviolet; a heavier pigmentation might do the trick under the more extreme Martian UV.

Of course, none of this suggests that we will find lichens or ants or cucumbers on Mars. Martian organisms, if they exist, will have followed entirely different evolutionary pathways. But it is interesting that even Earth life, having evolved under our mild conditions, is capable of surviving the Martian rigors. It does not prove that there *is* life on Mars. The fact that Mars is habitable does not necessarily mean that it is inhabited. But it certainly demonstrates that if life ever did arise there, it could easily have adapted to the harsh environment presented by that planet today.

However, under present Martian surface conditions, life could not possibly have arisen to begin with. The formation of life requires large bodies of liquid water, and some atmospheric constituents that do not exist on Mars today. So the question now is: was Mars ever different enough from its present condition to have allowed life to begin? If not, then all this speculation about adaptation and survival has been pointless. But if so, then the chances of finding Mars to be inhabited must be very good.

4

The Oceans of Mars

If Mars had an ocean, it would cover the northern hemisphere, and the southern hemisphere would be a single vast continent.

—*Thomas Mutch et al.*
The Geology of Mars

MARS TODAY resembles the frozen tundra of our polar regions. But evidence from the Viking mission has led most scientists to believe that it was not always so; that in the distant past, Mars had a climate and an atmosphere very much like our own, with warm temperatures, and an abundance of water flowing through innumerable rivers down to the shores of a vast ocean. White puffy clouds filled a deep blue sky, bringing rainfall to the now desertlike regions of the planet. Could life have flourished during this balmy, Earthlike period of Martian history? Certainly, if the scientists' reconstruction is correct, Mars could, in that era, have supported an extensive and diverse flora and fauna, perhaps even rivaling that of the Earth. In such a pleasant epoch, it would be no surprise at all to find that living beings, even highly evolved ones, had arisen there; on the contrary, it might be the most reasonable expectation.

How can we be so sure that this radically different climate

really did exist on Mars? Since the implications of such a past history are so momentous, let us review very carefully the evidence from which that history has been deduced.

The first and most obvious indication of a warmer past was shown clearly on the hundreds of Mariner 9 photographs that showed Mars to be covered by an extensive network of sinuous channels, so familiar in their resemblance to well-known terrestrial riverbeds that it was hard to remember that one was looking at pictures of another world, hundreds of millions of miles from our own.

Years of intense scientific debate have led right back to the most obvious conclusion: no reasonable alternative having been proposed, it is now almost universally agreed that these channels are, indeed, the dried up beds of thousands of ancient rivers that once drained almost every area of the Martian surface.[1]

This in itself was a breakthrough of gigantic import, for up until the discovery of riverbeds the strongest argument that had been advanced against the likelihood of life on Mars had been the supposed lack of water, which is one of the basic requirements for the existence of any kind of life that we know of: all of the fundamental biochemical processes that sustain the lives of human beings take place in a bath of water, which allows molecules to flow back and forth in the chemical activity that we call "metabolism." Some kind of liquid medium is essential for such chemical activity, and therefore for the existence of anything that we would call "life." Water seems like the ideal fluid medium, but it is not necessarily the only one: other possibilities include liquid ammonia, liquid methane, and liquid carbon dioxide, all of which are abundant on some planets. However, none of these is ever likely to have been available on Mars, so for our purposes the discovery of water there was an essential prerequisite for the possibility of life. And the implication of those riverbeds was not just that Mars had some water, but that it had vast quantities of it, as attested to by the number and size of the channels covering the face of Mars.

That is not the only thing that we can learn from those sinuous channels. Their existence proves not only the existence of water, but two other very important facts about what Mars used to be like, in terms of its temperature and its atmosphere.

Water simply cannot exist as a liquid on Mars today, and the fact that it flowed across the planet in the past means that the pressure of the atmosphere must have been much greater. The Martian atmospheric pressure at the present is less than one-hundredth of that on Earth, and in order for water to have flowed on Mars there must at one time have been at least twenty times the present amount.

A denser atmosphere is not enough, in itself, to permit water to flow on Mars. In addition, the temperature must have been substantially higher than it is at the present, otherwise the Martian water would simply have remained frozen.

These two effects—increases in temperature and pressure—go hand in hand because when the density of the atmosphere goes up, there is an increased "greenhouse effect": the greater amounts of carbon dioxide and/or water vapor in the air trap the sun's heat and cause a planetwide rise in temperature. Conversely, any increase in temperature melts additional ice and dry ice (frozen carbon dioxide) from Mars' polar regions, thereby increasing the concentration of these gases in the Martian atmosphere. Because of this spiraling effect, even a slight change in either the warmth or the air pressure on Mars may result in a drastic shift in its overall climate.

Such speculation is, of course, very indirect; the fact that such changes could occur does not establish that they did occur. Fortunately, these conclusions have been confirmed by an entirely independent line of reasoning based on highly sophisticated analysis of the Martian atmosphere by one of the Viking instruments.

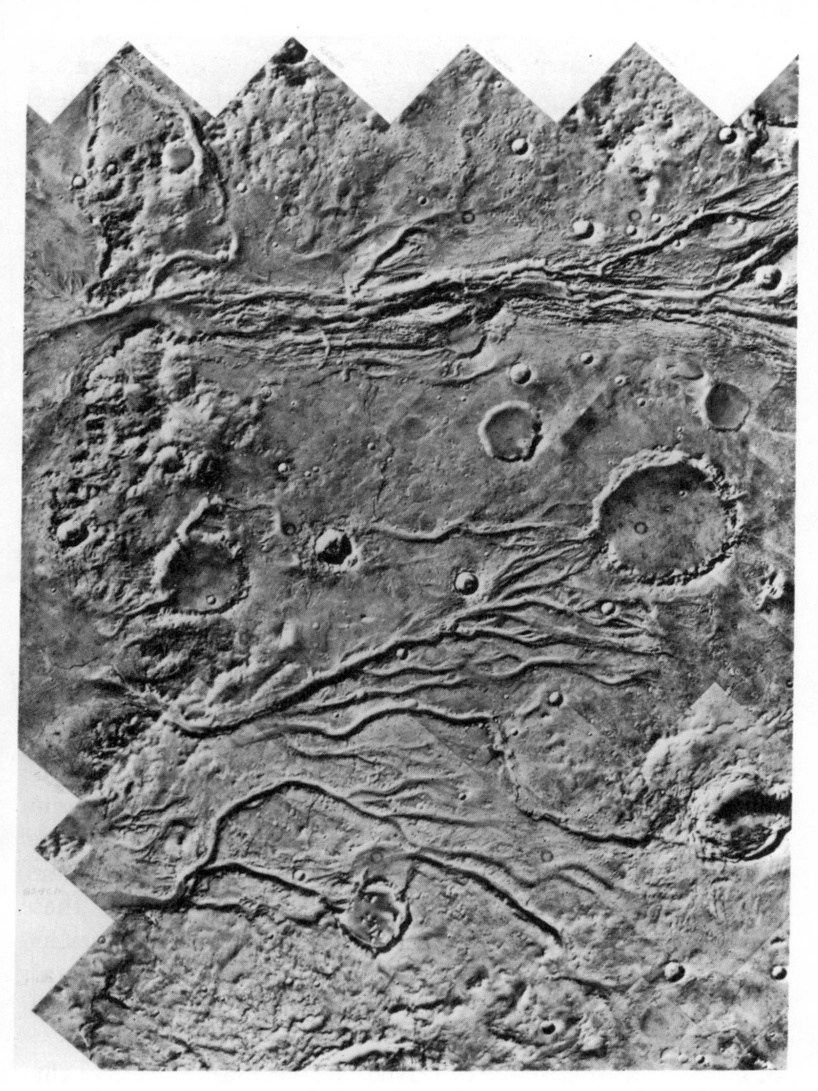

A sinuous channel, typical of those that cover most of the Martian surface. Scientists now agree that they must have been riverbeds during a warmer epoch of the Martian climate.

This instrument, the gas chromatograph mass spectrometer, was programmed to take measurements of the upper atmosphere as the lander made its descent to the surface. This device was capable of measuring not only the amounts of various gases in the atmosphere to a high degree of precision, but also the relative abundances of different isotopes of the same element.

Different isotopes of the same element, for example oxygen 16 and oxygen 18, are exactly identical to each other in every way except one: the higher-numbered isotopes are heavier, atom for atom, because they contain extra neutrons in the nucleus of each atom. Because these atoms are heavier, they are slower moving and are less likely to be disturbed by any dynamic process that is taking place. Thus, relative measurements of isotopes are an important clue to the history of the Martian atmosphere.

Since the Martian atmosphere is now very thin, it is dissipating at an appreciable rate into space. Since the lighter isotopes are more easily stirred up, they should dissipate more rapidly than the heavier isotopes. This would result in an increase in the relative abundance of heavy isotopes of each gas.

In the case of carbon dioxide and oxygen, the ratios that have been determined do not show this effect, that is, the ratios of heavy to light isotopes are about the same as they are on Earth. This means that there has to be some process that regularly renews the balance of the atmosphere by mixing it with a large volume of gas that has not been subject to this kind of selective dissipation.

In other words, in order to explain the observed ratios of isotopes in the present Martian atmosphere, it is necessary to assume that there are periods when the density of the atmosphere is much greater than it is now. The amount by which it must have increased can be precisely calculated on the basis of these measurements. This has been done by Drs.

Nier, McElroy, and Yung of Harvard University and the University of Minnesota, and their conclusion is that the atmospheric pressure that would be required to account for this data would have been twice the present pressure on Earth.[2]

The existence of Martian river channels requires that the atmosphere must at some time have been at least twenty times as dense as it is now. Precise measurements of isotopic ratios have shown that it may in fact have been more than two hundred times as dense. These two independent determinations leave no room for doubt that Mars has not always been as we see it today, with its thin, icy air. Sometime in its history, the air was thick and warm, and water was abundant.

If the atmospheric pressure really did equal or exceed our own, according to one calculation the resulting greenhouse effect might have raised the surface temperatures by 60 degrees or more,[3] bringing at least the tropical zones into a range that would have been comfortable for most Earth creatures, including humans. Furthermore, the denser atmosphere would have greatly smoothed out the daily variations in temperature, which now sometimes cover a range of almost 200 degrees, and brought them closer to some kind of equilibrium. In that era, Mars would have been a delightfully familiar and comfortable place to visit.

The Global Ice Age

How did a planet that once had a warm and pleasant climate, with water coursing in vast torrents through great meandering valleys, somehow end up as a frozen wasteland? The tracks left by thousands of years of abundantly flowing water, clearly etched on the Martian plains, prove that at some time the conditions there were drastically different. But how could such a catastrophic change in climate have come about?

One possible answer to that question can be found in

the history of our own planet. It is well known that the Earth has gone through long periods of extreme glaciation, the "ice ages," the most recent of which ended just 10,000 years ago. Recently, analysis of ocean sediments has proved that there were many more of these ice ages than scientists had thought.

This new information about the number and the exact periods of the ice ages has led to the revival of an old theory about their cause, a theory that until now had been considered farfetched and highly unlikely.

First proposed in the 1950s by Milutin Milankovitch, a Yugoslavian scientist, the theory holds that the cause of our cycle of glacial periods and warm intervals lies in the geometry of the Earth's rotation about the sun. Specifically, Milankovitch proposed that the ice ages might be caused by a combination of three different periodic changes: first, an oscillation of the Earth's orbit that causes it sometimes to be almost perfectly circular, and sometimes somewhat elliptical; second, an oscillation of the Earth's axial tilt, which is presently about 24 degrees from vertical but which varies from 22 to 25 degrees over a period of about 42,000 years; and third, the "precession of equinoxes," which, like the wobble of a spinning top, causes the Earth's tilted axis to point toward different parts of the sky and makes the spring and autumn equinoxes come slightly earlier each year, in a cycle that takes 19,000 years to complete.

At the time Milankovitch first proposed this theory, there simply wasn't enough data to either confirm it or refute it, and so it remained a subject of heated debate until a few years ago when detailed analyses began to be made of cores dug from the sediment on the ocean floor in many different locations around the world. These sediments, whose slow deposition over thousands of years left identifiable evidence of the world's changing temperature, supplied the missing data, and the result was startling: the detailed dating of the ice ages for the last 700,000 years matches perfectly the predictions of the Milankovitch theory. The theory had predicted seven major glaciations during that period, and all seven appeared at exactly the expected times. As a result of this close agreement, the theory is now almost universally accepted.[4]

What happened to the Earth during the ice ages was a small-scale version of what has happened to Mars: a large amount of the planet's water became frozen near the polar regions, and the level of the oceans became correspondingly lower. During the last ice age, the Earth's sea level was about 400 feet lower than it is now. It is easy to imagine that if the process of glaciation went further, the oceans might freeze up altogether, leaving the Earth as a frozen snowball without a trace of liquid water, just like Mars today. So it is entirely possible that Mars goes through a cycle of ice ages similar to ours, but much more extreme, and that we just happen to be seeing it now at its worst.

One possible explanation for the greater magnitude of the changes on Mars lies in the extreme eccentricity of its orbit, that is, the fact that at one end of its orbit, it lies almost thirty million miles closer to the sun than it does at the other extreme. At the present time, this closest point in its orbit coincides with the point at which its axis is tilted with the North pole away from the sun—the middle of winter in the Northern hemisphere. Therefore, the maximum warmth of the sun reaches the Martian surface in the Southern hemisphere, where it is then midsummer.

At the other end of its orbit, when summer reaches the Northern hemisphere, Mars is much farther from the sun, and so the Northern summer is not nearly as warm as the Southern. This is why the polar ice cap at the North pole of Mars is much larger than the one at the South pole, and why it never melts completely as the Southern cap does every year.

But this has not always been the case. As with the Earth,

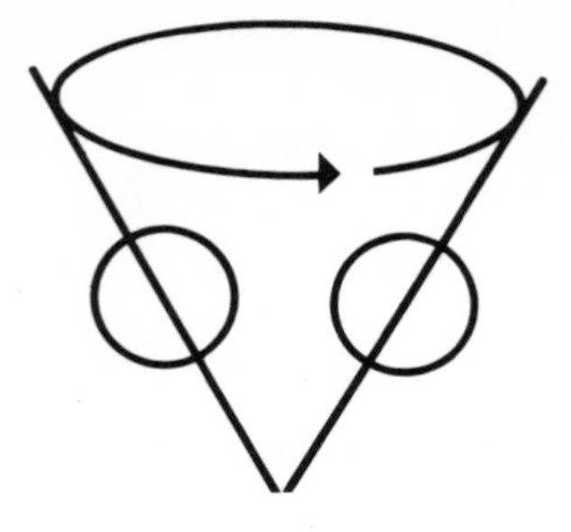

The Precession of the Equinoxes. Like a spinning top, Mars wobbles very slowly on its axis as it rotates. The wobble takes 25,000 years to bring Mars' axis through a full circle back to its starting point.

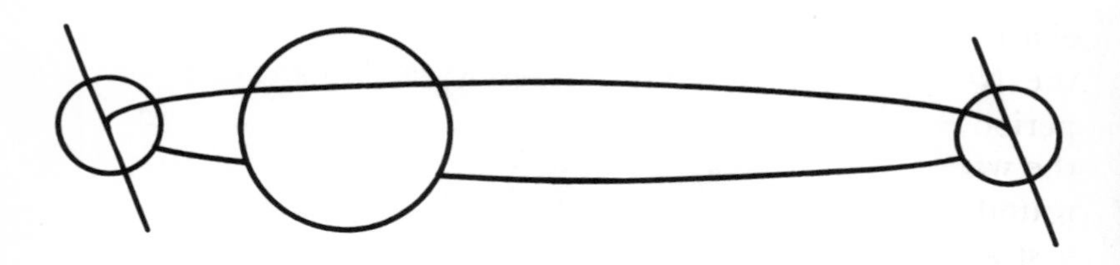

At present, Mars is closest to the sun when it is winter in the Northern hemisphere. By summer, when the Northern hemisphere faces the sun, it receives less sunlight because of the greater distance. That is why the northern polar ice cap never melts, unlike the southern cap which melts every summer.

12,500 years from now, at the other end of the recessional cycle, the situation will be reversed: the Northern hemisphere will be closest *to the sun during the summer, so its ice will melt every year. The Southern hemisphere will then have a permanent ice cap.*

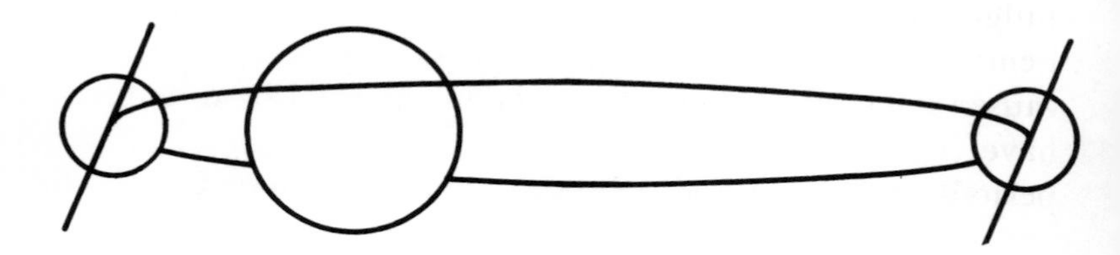

the direction of Mars' axial tilt changes, causing a precession of the equinoxes. As a result, the present situation will be exactly reversed in about 25,000 years; by that time, the *South* pole will be tilted away from the sun at the time of closest approach, and the North pole will get the benefit of greater warmth from the sun. The North polar cap will then melt, and it is the Southern one which will remain all year round.

But what happens in between? There will be a point, somewhere in the middle of this cycle, when there will be an exact balance between the amount of sunlight received by the Northern and Southern hemispheres. It is during this period when it is entirely possible that most, or even all, of the water from the polar ice caps may remain melted all year round; there might be a period of equilibrium in which a vast amount of water vapor would saturate the Martian air, and the climate would warm up—an "interglacial" period analogous to the one that the Earth is now in. And this warm Martian period might last for thousands of years.[5]

There are additional factors that might contribute to the effect: all of the variables involved in the Milankovich theory of the Earth's ice ages also apply to Mars, and in most cases they are much more extreme there. For example, the variation in the axial tilt, which on Earth changes by a scant 3 degrees, on Mars has a range from 14 degrees to 35 degrees, seven times the terrestrial range. It should therefore be no great surprise to find that the Martian climate goes through a more drastic cycle of changes than the Earth's.

Volcanic activity may also have played a part in the process: the intense volcanism that gave rise to the "Tharsis bulge," that enormous upwelling of land in the Northern hemisphere of which Mount Olympus is one part, may have caused an additional wobble in Mars' axial motion and may have contributed to the current cooling. Conversely, some periods of intense volcanic activity may have spewed large amounts of dark dust across the land, causing it to absorb more of the sun's heat and thus contributing to a warming of the planet. Also, volcanic eruptions put extra "volatiles," gases and water vapor, into the air, causing a temporary thickening of the atmosphere that could also contribute to an overall warming trend.

But such factors would presumably have been of secondary importance. It is reasonable to assume that since the Earth goes through great fluctuations in global climate, Mars would be subject to the same processes. Since the correlation of our ice ages with the mechanics of our orbital motion is now well established, that must also have been the dominant factor in Mars' climatic cycle. The one other factor that may have contributed significantly to temperature variations on both planets is the influence of the sun.

Recent studies have shown that the sun is not the constant and dependable source of heat and light that it was once believed to be. In fact, it seems that there may have been minor, but measurable, variations in solar output even within historical times: the "little ice age" that occurred in the seventeenth and eighteenth centuries, in which average temperatures in the temperate zones were several degrees cooler than their current averages, is now widely believed to have resulted from a decrease of a few percent in the sun's heat.[6] Such variations in the sun's radiation may themselves form a periodic cycle, but if so it is one that we have not yet been able to measure.

The net effect of all these influences may simply be to make the overall pattern somewhat erratic: some ice ages may be much more severe than others. As our exploration of Mars continues, important light may be shed on our own history: if we determine that Mars and the Earth have suffered from extreme cooling at precisely the same times in the past, then we will know that the sun's influence was the primary cause of those cooling events. If, on the other hand, we find that the Martian ice ages are precisely determined by the orbital variations of the planet, then that will have been firmly established as *the* cause, rather than simply one contributing cause, of our own ice ages. Either way, the discovery will be of great importance in allowing us to predict the future of the Earth's climate: if we know well in advance that another ice age is due, we may have time to prevent it—for instance, by putting additional carbon dioxide into the air and causing an increased greenhouse effect—or, at the very least, to prepare for it by shifting people and agriculture toward the south.

All of the foregoing relates to periodic changes—a cycle of ice ages and warm intervals. But can we be sure that climate changes on Mars have in fact been a recurring phenomenon?

Many scientists have held that this was *not* the case; they believed that Mars may have had a warmer climate for one brief spell several billion years ago, but that it has been in its current icy state ever since. However, two kinds of evidence from the Viking mission have thoroughly refuted that contention.

First of all, the layered pattern of the North polar ice demonstrates very graphically a history of melting and refreezing that must span hundreds of millions of years. This layering fits very nicely with the theory that the precessional cycle, with its alternate melting of the North and South polar ice caps, may have dominated the Martian climate.

But the most compelling evidence, which has completely taken the wind from the sails of those who supported the "steady decline" theory of Martian climate, comes from a study of the river channels. This study, conducted by Harold Masursky and his colleagues at the U.S. Geological Survey, was an attempt to determine the ages of some of the channels by using crater counts: the older areas of the Martian surface contain a larger number of craters in a given area than areas that have more recently undergone geological changes through volcanic action or through erosion. Thus, by counting the number of craters in and around various channels, a rough idea of their relative ages can be determined. This method is not precise enough to give an exact dating of the Martian warm periods, but it has proved that there have been many such episodes: the conclusion from this study was that "we have documented channel ages (by crater counts) that span the entire time interval that can be analyzed." [7] That time interval covers three billion years, from 3.5 billion years ago (about the time life began on Earth) to less than half a billion years ago. This clearly establishes that Mars has experienced warm intervals throughout its history.

This same study also established that the river channels really were caused by rainfall. Some geologists had held that the channels resulted from the melting of subsurface ice, which broke through the surface in a sudden flood. If that had been the case, then it might have meant that Mars never did get as warm as we had thought. The melting might have come from volcanic heat and caused a very temporary appearance of some liquid water on a planet that might not have been very much warmer than it is now—certainly not warm enough to evaporate water and cause rain to fall.

But such arguments have now been put to rest. Masursky's study has clearly shown that many of the channels have the characteristic "dendritic" (branching, or treelike) patterns associated with the collection of rainfall over a wide area, the familiar pattern of ever-smaller tributaries that we see in terrestrial river systems. There are none of the features associated with the melting of ground ice: slumping, chaotic terrain, box canyons, and so forth. It is clear that at least some, if not all, of the Martian channels were in fact caused by rainfall.

The Martian Oceans

There are three components of the well-known "water cycle" on Earth: it begins with the ocean, where the sun's heat evaporates water into the atmosphere; the resulting clouds are blown over the land by winds, where they eventually lose their moisture in the form of rainfall; the rain is caught in the drainage basin of a river system, from which it eventually flows back into the sea, where the cycle begins again. We now know for sure that two of these three elements existed on Mars: rainfall and rivers. Might there also have been oceans?

On the face of it, one would assume so. Oceans would fill the "missing link" in the Martian water cycle. After all, rivers do not just disappear into thin air, and the presence of rivers would seem, in itself, to be strong evidence for seas or oceans into which they might have flowed. As it turns out, there is considerable other evidence to support this conclusion.

The first set of supporting evidence comes from the topography itself. A large number of the major channels flow toward low-lying areas. These lowlands tend to be surrounded by a perimeter of steep cliffs which closely resemble the edges of our own continental shelves. These cliffs, which are routinely referred to as a "continental margin" by geologists such as Harold Masursky,[8] reveal not only a sudden change in elevation but also the probability of long-term erosion such as might be expected at the shore of an ocean. Since most of the low-lying areas of Mars are bordered by what seem to be familiar kinds of coastal features, it is reasonable to think that these regions were, in fact, oceans during the warm epochs of the Martian climate.

This is supported by a comparison of the highland areas with the former oceanic floor. The highlands are covered with large numbers of craters, almost to the point of "saturation" found on the moon (at which point each new crater formed would, on the average, obliterate a previously existing crater, so that the overall number of craters remains essentially constant once saturation is reached). The lowlands, however, are relatively smooth and uncratered; there are some craters in these regions, but considerably fewer than on the highland or "continental" areas. This is just what we might have expected, since the sea-floor would have experienced strong erosional forces, as well as the accumulation of sediments, which would have obliterated or buried most of the craters that were once there.

The boundary between the plains and the cratered highlands has been mapped in some detail by Thomas Mutch and his associates in their book "The Geology of Mars," the most exhaustive text on the subject published so far.[9] Although they do not believe that a Martian ocean ever existed, they have done a splendid job of analyzing exactly where it would have been. Their maps allow a clear picture to be drawn of just what Mars must have looked like if and when the ocean was there.

The basic picture, as expressed in the quote at the beginning of this chapter, is of a single great continent in the Southern hemisphere and a vast ocean occupying most of the Northern hemisphere. Actually, the picture is slightly more complex than that. The boundary between ocean and continent does not follow the equator, but rather is tilted by about 35 degrees from the equator. Also, there are some highland regions, notably the volcanic areas of the Tharsis bulge and of Elysium, which extend into the northern oceanic area. There are several large basins, notably Hellas and Argyre, which show all the features of the oceanic areas—smooth, low-lying plains surrounded by a steep escarpment—and which must therefore have been smaller landlocked seas near the South pole of Mars. The North pole of Mars is an elevated area, so apparently there would have been a small polar continent in the northern ocean whose size and position closely resemble Antarctica.

According to Mutch, "The margin between continent and ocean would be marked by a sharp declivity, which drops two or three kilometers over a distance of several hundred kilometers. The scale of this slope is similar to the slopes between ocean basins and continents on Earth." [10]

In short, from a topographic and geological standpoint, Mars *looks* as though it has had an ocean. In itself, this evidence might not be wholly convincing, although it is hard to imagine how the clearly visible coastal features, which even the skeptical Mutch describes as an "erosional contact," [11] could possibly have formed by any process other than long-term erosion by a large body of water.

But there is additional evidence. An investigation of the orbiter photographs by a team of Viking scientists has discovered that there is a type of crater found on Mars that is totally unlike any of the craters on the moon or Mercury—the two heavily-cratered, lifeless worlds that have been photographed and studied in detail. These unusual Martian craters are characterized by a large pattern of material that appears to have flowed out from the point of impact, and the heavy, viscous pattern of flow suggests that the flowing material had approximately the consistency of mud.[12]

By contrast, there is no flow at all in lunar and Mercurian craters. The material thrown out of the crater is all in separate, solid chunks and the result is a pattern of debris strewn randomly around the crater wall. Some of the Martian craters are also of this type, and there seems to be a regional distinction. Some areas of Mars have only the "wet," flowing type of crater, other regions have only the "dry," brittle type. Since the Viking orbiters have so far only photographed a tiny fraction of the planet's surface, no large scale analysis of the distribution of the two different kinds of craters has yet been undertaken. But even a brief look at the photographs that have so far been analyzed makes the pattern strikingly apparent: *all* of the craters found in the oceanic areas are of the wet, flowing type, and all of the craters found in the highland, continental areas are of the dry, lunar type.[13]

What caused the "flowing" pattern of these craters? The

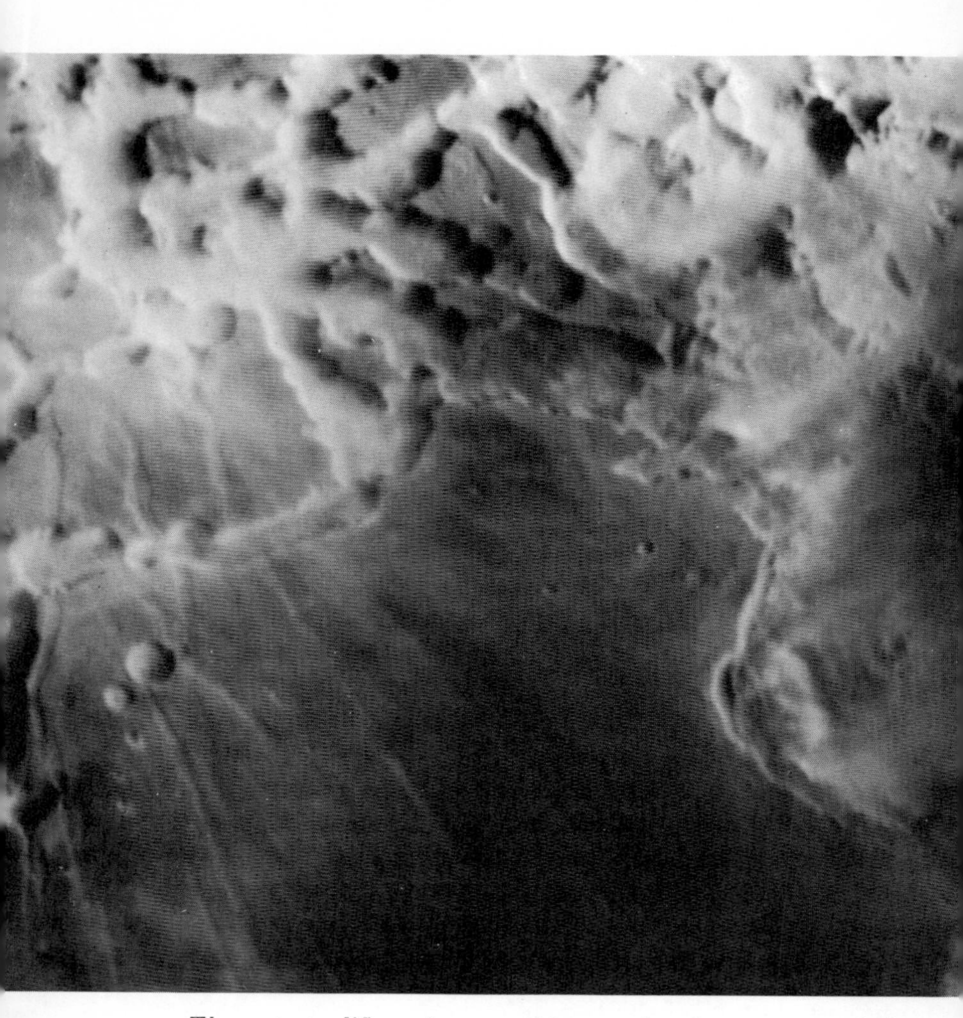

These steep cliffs and mesas appear to be the remnants of the coastline of an ancient Martian ocean. Fog can be seen in the low-lying areas in this early morning photograph.

Viking team explored a number of possibilities, and concluded that by far the most likely explanation is that these regions contain a large proportion of subsurface ice. When a meteor struck the surface, the heat generated by the impact melted the ice and created a large pool of mud which then flowed outward in a great, gooey mass until it cooled off enough to freeze solid in its distinctive, flowing shape.

The fact that these flowing craters are found exclusively in the oceanic areas proves that these regions contain a vast reservoir of ice just below the surface. This had been predicted by geologist D. M. Anderson, who had suggested in 1973 that (here paraphrased by Thomas Mutch) "in an earlier stage of Martian development, large amounts of water may have formed surface water, possibly on the scale of a large sea. As the atmosphere cooled, this sea may have been transformed to a vast cake of ice and buried under eolian [windblown] sediments." [14]

The freezing of a large ocean seems to be the only plausible mechanism whereby extensive sheets of ice could have formed under such large areas of the Martian surface. Since the evidence for the existence of the subsurface ice is quite clear cut, the case for a Martian ocean is greatly strengthened.

The clincher was provided by the Viking landers. Both of the landing sites were selected for a similar set of characteristics, which included the requirement that they be in smooth, low-lying areas. So it is not by coincidence but by design that both of the landers ended up on the floor of the ancient Martian ocean. And, looking at the photographs taken by the landers, it is easy to imagine that one is looking at an ocean floor from which the tide has temporarily receded. Anyone who has seen the Bay of Fundy, with its unusual tides, will find much that is familiar in these boulder-strewn Viking landscapes.

Furthermore, when the Viking sampler arm dug into the surface, the resulting trench did not collapse like loose

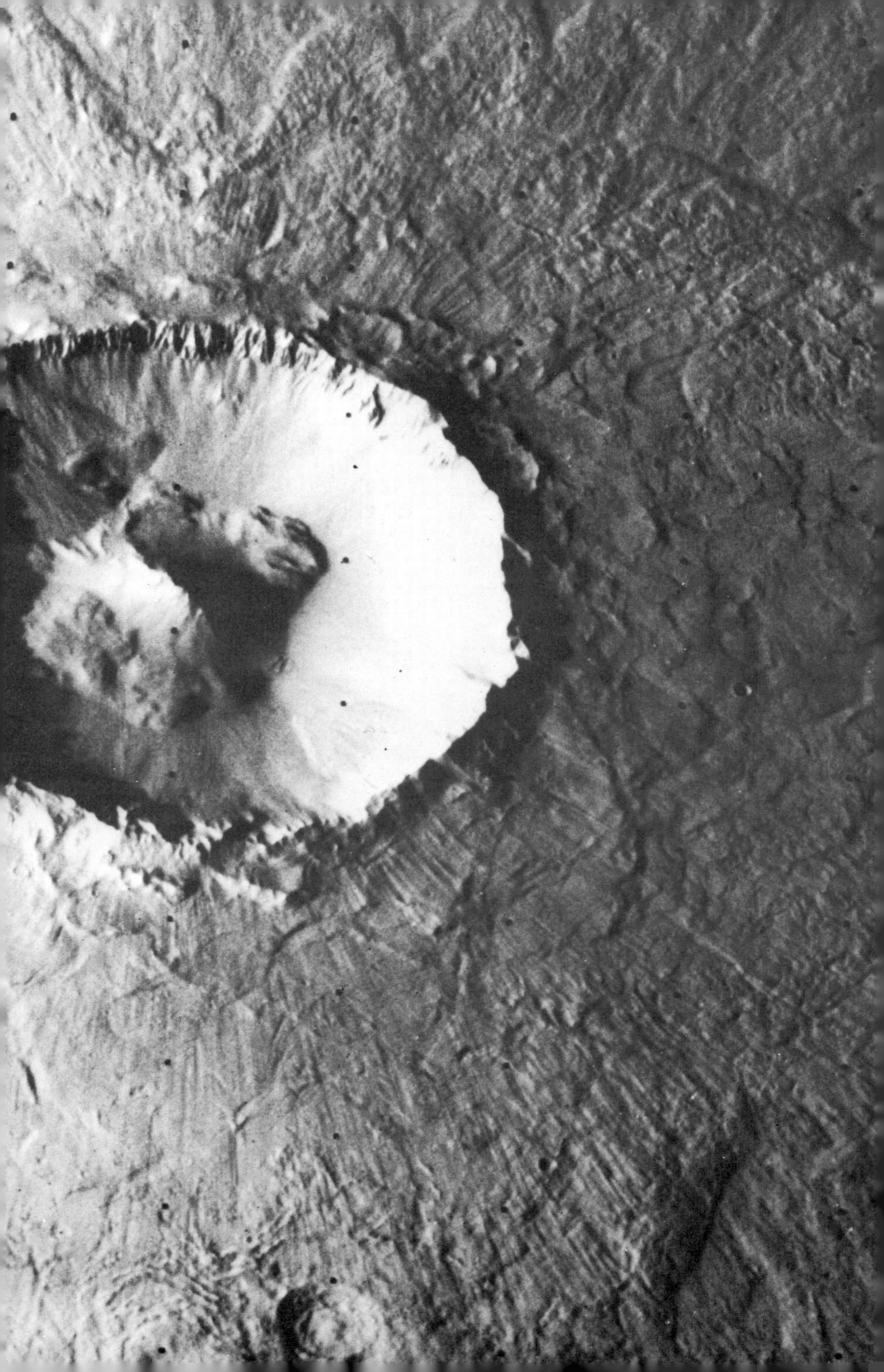

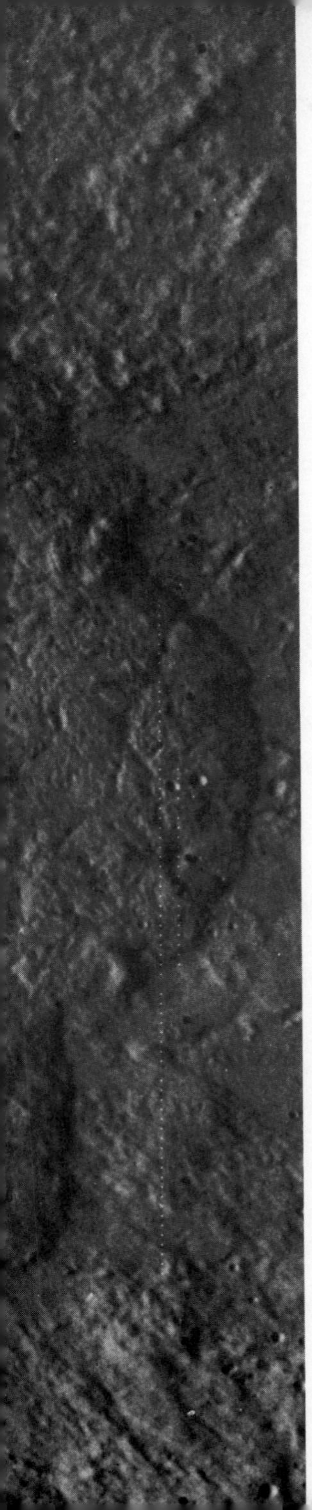

This "flowing" crater type is unique to Mars, and probably results from the melting of subsurface ice by the heat of impact of a large meteorite. The mud thus formed flowed outward from the crater until it cooled off enough to freeze in place, preserving its flowing shape.

soil, but kept its hard-edged appearance in a way that reminded the experimenters of wet sand. The clear implication of this cohesion of the soil is that it must have a very high water content—in short, the ground appeared to be frozen solid.

But it is not the appearance of the landing sites that gives such strong support to the theory; it is the results of their chemical experiments on the Martian soil. The soil at both sites is almost identical, and according to the chemical analysis conducted by the X-ray fluorescence spectrometer, it seems to be a kind of clay very rich in iron oxides. In fact, according to a report by the Viking geological team, the closest parallel known on Earth to the soil found on Mars is the mud at the bottom of the Red Sea.[15]

So we seem to have a remarkable convergence of four kinds of evidence: the coastal features at the edge of the lowlands, the smoothness of these areas, the "wet craters," and now a direct chemical analysis of the soil, all of which are consistent with the features that one would expect to find associated with a dried up or frozen over ocean basin.

Can this be just some miraculous coincidence? Surely it is more reasonable to conclude that all those rivers whose mouths run into the lowland areas really were feeding a large, globe-circling ocean, which must have made Mars, at those times in its past, seem very Earthlike indeed: we might have been able to stand on a Martian beach watching the deep blue ocean stretching as far as the eye could see, the surf crashing against giant stone cliffs, a mighty river flowing into the sea beneath a blue sky dotted with woolly clouds. The temperature may have been similar to that of a nice sunny beach day in summertime on Earth, with a cool breeze blowing briskly off the ocean. The only visible difference might have been the pale yellow sun, shining only half as brightly as it does on Earth. Were it not for that one difference, a person who found himself magically transported to this long-ago Mars might not have been able to tell that he was not on some unknown shore right here on Earth. Our sister planet may have been very nearly our twin.

The Rusty Planet

There is one further bit of evidence for all this speculation, and it leads to the most exciting conclusion of all. It is based on Mars' most obvious attribute. The most apparent thing about Mars is its redness. This coppery hue is clearly visible to the naked eye, and makes it easy to distinguish "the red planet" from all the other stars and planets in the heavens. The vividness of this color was made even more apparent by the Viking lander photographs, which showed a surface that looked very much like an expanse of rusty iron.

After careful investigation, geologists have decided that Mars is covered by a layer of rusty iron, or iron oxide. Specifically, they believe a large amount of the surface material is made up of an iron oxide called limonite, which is common in some desert regions in the Earth's tropics.

Aside from explaining Mars' vivid coloration, which was the earliest observed characteristic of the planet (having been known for millennia), this abundance of limonite reveals some very significant details of Martian geological history. Limonite on Earth only arises from a combination of extreme humidity (which is why it occurs almost exclusively in the tropics) and the abundant oxygen in our air. The reason for this is obvious from its chemical composition: a molecule of limonite consists of two atoms of iron and three of oxygen, attached to one or more water molecules (this is written as Fe_2O_3-nH_2O). The presence of water and oxygen is clearly an absolute prerequisite for its formation. The Martian air must have once contained substantial amounts of oxygen and water vapor. (The quantity had to be substantial, in order to have rusted the entire planet's surface down to a depth of at least a few inches.) The water vapor fits in nicely with all the other evidence for a past warm climate on Mars with oceans and abundant rainfall, all of which contributes to the likelihood that life may have appeared on Mars at some time in the past. But the evidence for oxygen is even more significant.

Oxygen, as far as geologists have been able to determine, only came to exist in Earth's atmosphere as a result of photosynthesis, the action of living plants. In fact, no one has been

This map by the author, based on U.S. Geological Survey topographic maps of Mars and on the erosional evidence given in The

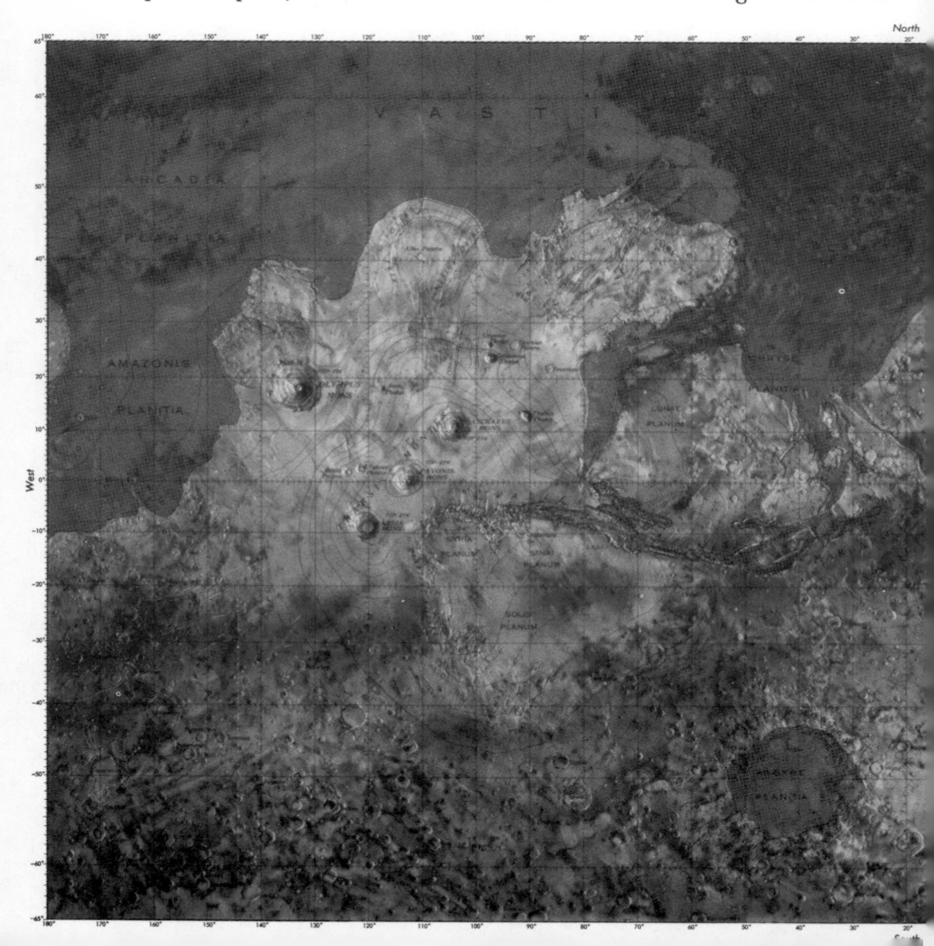

able to think of any other possible way for oxygen ever to accumulate in a planet's atmosphere. If all life on Earth were to disappear, all of the oxygen would be gone in a very short time. This has often been suggested as a surefire way of discovering the existence of life on another world: if we ever find a planet that has any significant amount of oxygen in its atmosphere, most biologists would take that as proof positive that life exists there.

Geology of Mars, *shows how the planet may have looked during the era (or eras) of the Martian ocean.* U.S. Geological Survey.

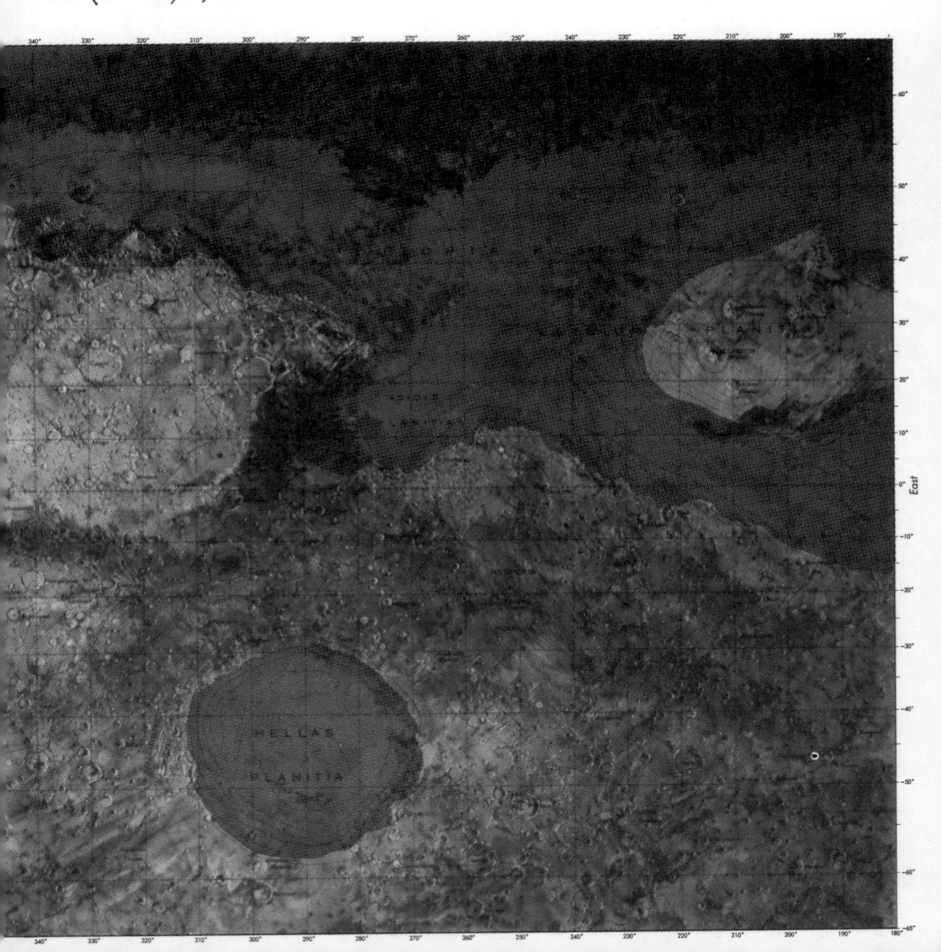

There is no oxygen to speak of in the Martian atmosphere today, so we do not have that proof. But the silent testimony of all that rusty iron strongly suggests that an oxygen-rich atmosphere once existed there, and so implies

This map of the Chryse Planitia area of Mars shows part of the basin that may once have been an ocean. The white X (upper right) marks the location of the Viking I lander. This is a U.S. Geological

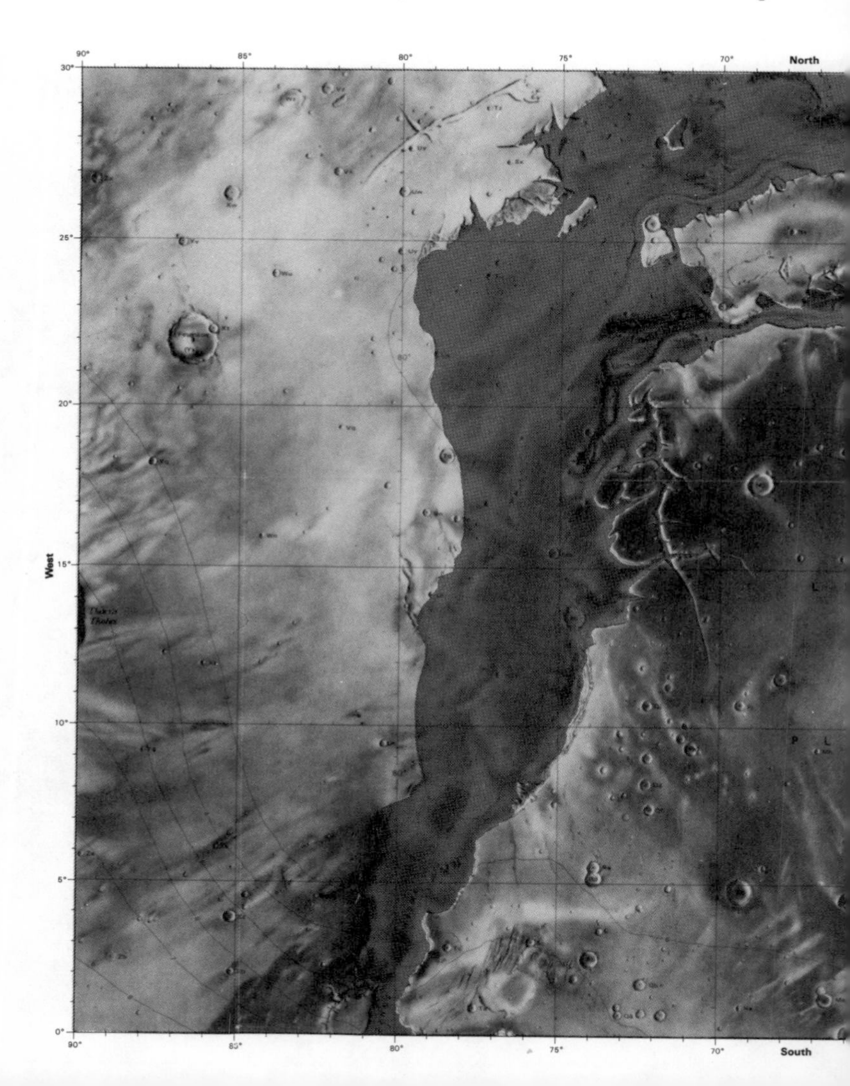

that vegetation, and a very extensive and thriving vegetation at that, must have flourished in some ancient Martian era.[16]

It seems that the red planet may once have been green.

Survey map, with the ocean area—based on topographic and geological information—added by the author. U.S. Geological Survey.

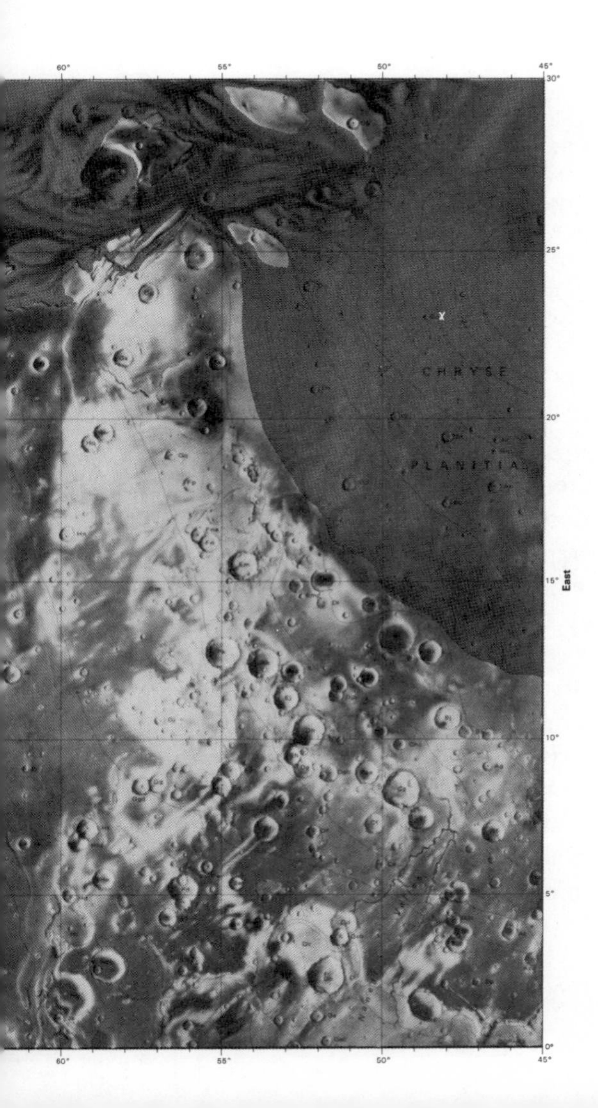

The Viking Search for Life

An unambiguous resolution of the question of Martian life may not be obtained for decades. My view of the evidence is that although any single test result could have a chemical explanation, when all the tests are considered together they suggest that life or some process imitating life exists on Mars today.

—*Robert Jastrow*
Until the Sun Dies

THE STAGE has been set. We have looked at Mars closely, and we know that there is no reason not to expect to find it inhabited. All the necessary prerequisites for life are there; the time has come to find out whether the potential has been realized, whether the habitable planet is, in fact, inhabited.

If this were "Star Trek," there would be nothing to it. Mr. Spock would point a tricorder at the alien planet, and in two seconds flat he would read the results right off the screen: "Well, Captain, it seems that we have primitive life forms on this planet, including plants and insect life, but no intelligent beings." Depending on their mission, they might beam down to sample the vegetation, or they might go on their way, searching for planets populated by strange aliens capable of embroiling them in more interesting plot lines.

Alas, the detection of alien life is not that simple. There is no single test, or even series of tests, that can identify "life" in any of the myriad forms that it might conceivably take. Life is a very complicated process—indeed, the most complex that we know of in the universe—and it is hard for scientists to arrive at a clear-cut, all-embracing *definition* of life, much less an unambiguous test for it.

That is why there has been a great deal of confusion and misinformation about the results of the Viking biology experiments. People had hoped that after the soil was dumped in the hopper, a computer display would light up to proclaim "Yes, there is life on Mars" or "No, there is not." If only that were possible!

In fact, as the scientists who had been planning these experiments for almost a decade knew perfectly well, such certainty was extremely unlikely. Everyone agreed that there could be no such thing as a negative answer. Even a clear, unambiguously negative result from every one of the biology experiments would not have proved that Mars was uninhabited. The tests were much too limited to allow such a sweeping conclusion; we might just have been looking in the wrong place, or at the wrong time of year, or using the wrong instruments.

According to Harold Klein, Viking chief biologist, looking for life on Mars is a bit like going fishing in an unknown lake. He says. "We are fishing in a lake in which nobody knows if there are any fish or, if there are any fish, exactly what they may be like." [1] Under the circumstances, any fishing you do will be, at best, a gamble.

The strategy adopted by the Viking biologists was to use three entirely different kinds of bait, in the hope that one of them might turn out, by a stroke of luck, to be just the thing that Martian creatures like to nibble on. So, a positive response to any *one* of the three experiments would mean that they had guessed right, and that there really is life on

Mars. But even a negative result, according to Dr. Klein, "may mean that there is a biology, but that we are using the wrong bait to find it." [2]

The results turned out to be harder to interpret than they had anticipated. Two of the experiments did yield results that met all the criteria, agreed upon before the launch, for a "presumptive positive." [3] Both of these experiments also met the next test, which, in the original conception of the experiments, had been thought of as the clincher for a yes verdict on Martian biology: when they were repeated as control experiments by first sterilizing the soil with intense heat, the reactions were eliminated. It certainly appeared that whatever organisms had been causing the reactions were killed off by the heat.

In the first few days after these results came in, there was a mood of intense excitement at Viking headquarters. Most of the scientists who had planned these biology experiments had never really expected to get a positive result. The discovery of life on Mars had been considered, at best, a long shot. But now, here were these incredibly strong positive results coming back, suggesting that maybe there really *was* life out there, after all.

This stunning realization had a sobering effect. Suddenly, it was clear that this was no longer a simple chemical experiment. It was clear that the Viking scientists were in the middle of what might turn out to be the most important discovery of the twentieth century: the first discovery of extraterrestrial life. Suddenly, the scientists realized the weight of the decision they were trying to make. If they were to declare that they had really discovered life on Mars, the world would never be quite the same. Scientific efforts for the next few decades would be geared, above all, to finding out more about this new form of life. Our feelings, as a species, about our place in the universe would be forever altered. This was not a decision that could be made lightly.

It became clear that the duty of the Viking scientists was to remain as cautious and as conservative as possible in interpreting their results. They simply could not afford to make a pronouncement about the existence of life on Mars if there was any chance at all, however remote, that they might be wrong.

As Dr. Klein put it, "Life is the most complex peak of evolution that we have seen. Therefore, we must try every other possibility to explain the responses by physical means, by chemical means, before being absolutely driven, you might say, to the conclusion that we can only explain it by a living reaction." [4]

In most areas of life, we do not even strive for such absolute proof. Even in convicting someone of a serious crime, a court demands only proof "beyond a reasonable doubt." That is a very subjective decision, and the determination of how much doubt is "reasonable" is left to the agreement of twelve individuals. But in order to proclaim the existence of life on Mars, the scientists will not settle for anything less than *absolute* proof. And that, it is now generally agreed, could not be obtained from the Viking experiments—very strong evidence, yes; absolute proof, no.

Unfortunately, because of the necessary caution of the Viking scientists, most people are not aware of just how close we *are* to the proof of life on Mars. In the newspaper and television reports, it often appeared that the experiments were hopelessly inconclusive.

Nothing could be farther from the truth. The results were stunningly positive, more so than anyone expected. Dr. Robert Jastrow, who has been the director of NASA's Goddard Institute for Space Studies for more than twenty years, says that: "Short of seeing something wiggling on the end of a pin, the case for life on Mars is now as complete as the Viking experiments could make it."

And in his summary of the biology experiments in the

final Viking scientific report, biology team leader Dr. Harold Klein concludes that: "On the basis of all the experiments performed to date, the Labeled Release experiment, unlike the other biological experiments, yielded data which met the criteria originally developed for a positive. On this basis alone the conclusion would have to be drawn that metabolizing organisms were indeed present in all samples tested." [5] He adds that it is possible that an unusual chemical reaction might have caused the results, but he feels that: "It certainly leaves the question of biology open. Of the three fishing lines, this one seems to have the best bait. [But] we have limited ourselves to three fishing lines and we should have had a thousand for such a momentous task." [6]

The Viking Experiments

The Viking biology laboratories are very sophisticated machines, containing the most advanced set of remote-controlled instruments ever assembled. The three experiments in these robot laboratories were brilliantly conceived to provide a clear indication of the presence of life processes, even though the processes might be unfamiliar.

All three of the experiments were designed to look for microbes, living things too small to be seen with the naked eye, such as bacteria, algae, and protozoa. There are two reasons why the search was directed toward microbes: First, practical constraints on the size and weight of the detection equipment dictated that the experiments had to be done with a tiny amount of soil, too small for anything larger than a microbe. Second, microbes were the first living things to evolve, and they survive in environments where no other living things are found, such as the dry valleys of Antarctica and the deepest parts of the oceans. So, if there were any kind of life at all on Mars, it would presumably include microbial life.

Despite the incredible variety of microbes, some universal characteristics are now known which clearly distinguish them from nonliving matter. All of them go through some kind of metabolism, that is, they ingest certain chemical substances, break them down and rearrange them chemically, and then release by-products, usually as a gas. For example, every marshy area contains billions of bacteria that eat decaying plant matter and release methane gas.

All three of the Viking biology experiments were designed to detect parts of this process: one of them, by measuring the assimilation of gases from the air into microbes in the soil; the other two, by feeding nutrients to the soil and measuring any gases that were given off.

The Pyrolitic Release Experiment

Evidence that strongly suggested the presence of living microbes in the soil was obtained from the pyrolitic release test, in which a soil sample was placed in an environment that closely duplicated normal Martian conditions. The results seemed to indicate that there were plantlike organisms in the sample, perhaps resembling terrestrial algae.

In this experiment, a spoonful of soil was sealed in a small cylindrical chamber along with some Martian air (which is mostly carbon dioxide). Then some radioactively-labeled carbon dioxide was added. The labeling, in which radioactive carbon 14 atoms were substituted for ordinary carbon, would not change the chemical properties of the gas; it was just a way of identifying those molecules that had been added, so that if they underwent chemical reactions, the labels would show up in the products of those reactions.

After the chamber was sealed, a xenon lamp was switched on to simulate Martian sunlight. Under those conditions, algae and all other earthly plant life would undergo the process of photosynthesis—the absorption of carbon dioxide from the air, and its conversion into complex organic molecules such as carbohydrates and sugars.

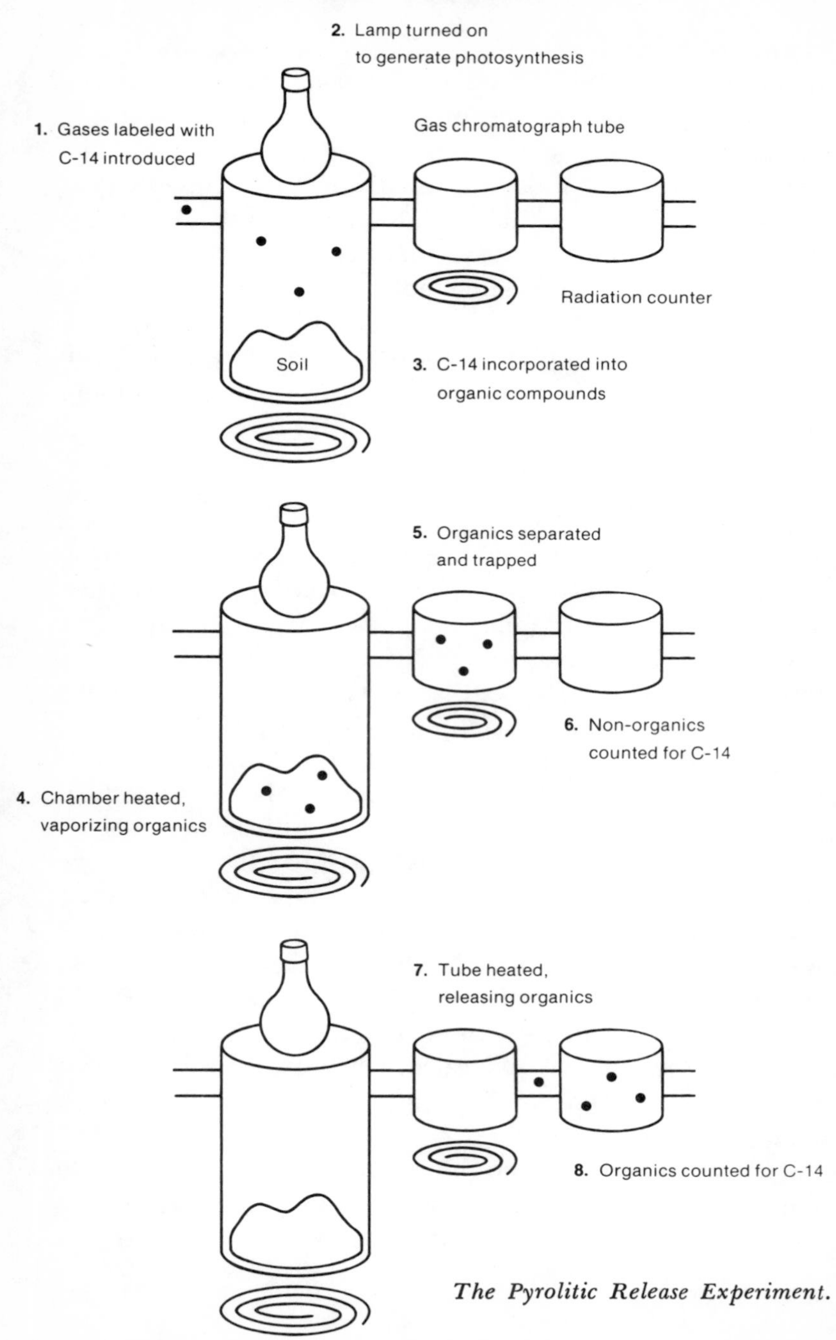

The Pyrolitic Release Experiment.

After a five-day incubation period, the chamber was heated to a temperature of 1,200 degrees, at which the cells of any organisms that may have been present in the soil would be broken down and vaporized.

The gases thus produced were then fed through a gas-chromatograph tube, which was filled with material that absorbs organic molecules while allowing inorganic molecules to pass freely. By heating the tube, the trapped organic molecules were then released into a radiation counter, which could detect the presence of any of the radioactively-labeled carbon atoms.

So, this experiment could only detect organic molecules —the complex compounds which are the basis of all known biology—and only those organic molecules which had actually been *created* during the incubation period. (Inorganic molecules would have been dissipated during the first heating, and organic molecules that were already in the soil before the test began would not be radioactive and so would not be detected).

In short, the amount of radioactivity detected by the counter would provide a direct quantitative indication of how much photosynthesis had taken place, that is, of how much carbon dioxide from the air was converted into organics in the soil.

If there was no photosynthesis, the detector would still register a background reading of fifteen counts, which thus corresponds to a zero result. This is the reading that was obtained in tests with sterilized soil, and with lifeless soil from the moon.[7]

But the very first sample of Martian soil produced a reading of ninety-six counts—more than a 500 percent increase over the background level. This was a clear-cut positive response, equal to readings obtained from Earth soil containing about 3,000 microbes.[8] No such result has ever been obtained from soil that did not contain living organisms.

In order to confirm the biological nature of this response, two kinds of control experiments were performed. In the first of these, the Martian soil was sterilized by heating it for three hours to a temperature of 340 degrees, which would kill anything that might have been living in the soil. (A friend of mine has described this procedure as "the first interplanetary murder.")

After sterilization, the count from the radiation detector was exactly the background level of fifteen. Apparently, the organisms responsible for the initial reaction had been killed off. According to the criteria established before the launch, all of the conditions for a "presumptive positive" had been satisfied. Apparently, life had been detected.

A subsequent control run was performed with a "cold sterilization" at a temperature of 90 degrees for two hours. The result from this run was positive; the reaction was unimpeded by the lower temperature.[9] While some have taken this as evidence against a biological interpretation, the fact is that many terrestrial microorganisms can easily survive such temperatures. In the harsh Martian environment, we might well expect to find unusually hardy organisms, whose survival under extremes of temperature would be no great surprise.

The other control experiment was a "dark run," in which an unsterilized sample was incubated with the light switched off. Since photosynthesis depends on light, the dark run should have produced a negative result if photosynthesis had been responsible for the original results. That is exactly what happened: the reading from the dark run was twenty-one, barely above the background level and within the margin of error. This negative result gave added support to the conclusion that the responses had been produced by photosynthetic organisms.

"At this point we got very excited," Dr. Klein recalls. "It certainly fit the concept that there was some biology going on." [10]

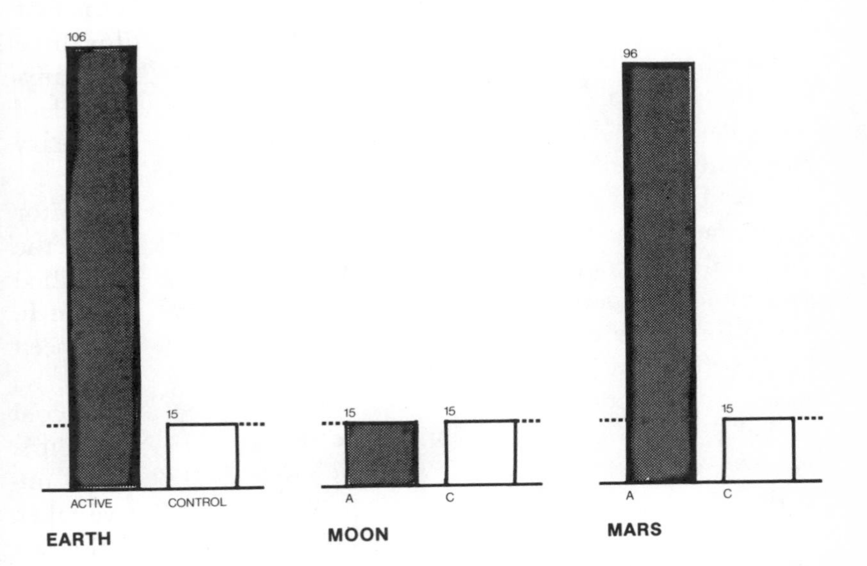

Pyrolitic Release Experiment Results. These bar graphs show the results of the pyrolitic release test during the active (shaded) and control (white) runs; the height of the bars indicates the number of counts of radioactivity per second, which should correspond to the amount of carbon dioxide assimilated by organisms in the soil. The results shown for the first Viking I sample (Mars) were almost identical to the results obtained from some Antarctic soil (Earth). The results from lunar soil (moon), known to be lifeless, are clearly negative. Mars thus seems much more Earthlike than moonlike, in this sensitive test for life.

But Dr. Norman Horowitz, who designed the experiment, is not yet convinced that it shows the presence of life. Always the most skeptical of the Viking scientists (before the mission began, he told a writer for *The New Yorker* magazine that the probability of ever finding any kind of life on Mars was a flat "zero"),[11] Horowitz still thinks that some unanticipated chemical reaction might account for the results.

And yet, as of this writing, nearly two years of intense experimentation by Horowitz and dozens of other biologists have failed to come up with any such chemical reaction. So far, biological activity is the only hypothesis that can account for the observed results.

Horowitz does concede that "it is not easy to point to a nonbiological explanation for the positive results," [12] and that "until a definitive explanation of the observations is forthcoming, a biological reaction will remain as a remote possibility." [13] Considering that just a few months before he had put the chances at zero, this statement may be revelatory of the direction in which the evidence is pointing.

The Gas Exchange Experiment

The gas exchange experiment was the only one of the three biology tests whose results were generally agreed to be negative. This is not especially surprising, since it was the experiment that deviated most radically from normal Martian conditions. Since Martian organisms are presumably adapted to Martian conditions, these changes, including higher temperature and much greater humidity than would ever be encountered on Mars, might be expected to affect them adversely. In fact, Horowitz had predicted before the experiments began that "if there are any organisms on Mars, they will surely drown or burst in Oyama's pharmacy." [14] (The experiment was designed by Dr. Vance Oyama.)

In this experiment, the soil sample was fed a nutrient

solution containing a wide variety of compounds believed to be desirable to almost any known organism. The solution included carbohydrates, proteins, and vitamins; while a person could probably have eaten it and derived some nutrition from it, it had a foul smell and would probably have caused a bad case of heartburn. The scientists referred to this rich nutritive broth as "chicken soup."

When the soup was added to the test chamber, the air in the chamber was monitored for changes in the concentration of any of several gases.[15] To the amazement of the experimenters, who had expected that if there was any reaction at all it would occur slowly and gradually, what they saw was instead a sudden, intense outpouring of oxygen. At first, this seemed like a strongly positive result. In his initial announcement to the press, Viking chief biologist Dr. Harold Klein said: "It is a stronger response than we have seen in fairly rich terrestrial soil, and it would imply that microbial life on Mars is highly developed—more intense than it is on Earth." [16]

But the reaction slowed to a standstill within two days, instead of continuing to accelerate as would a biological response. While this may mean that, as Horowitz had predicted, the organisms had burned themselves out in the heat and humidity of the test, most scientists are now convinced that the oxygen resulted from a purely chemical reaction, probably one involving a highly reactive agent such as a peroxide, ozonide, or superoxide.[17]

But this reaction, like the previous one, was eliminated by sterilizing the soil, so a biological interpretation of the result cannot be ruled out.

The Labeled Release Experiment

The strongest evidence for life came from the labeled release experiment which, like the gas exchange experiment, used a nutrient solution. The difference was that in this ex-

GAS EXCHANGE

Gas detector

Nutrient

Soil

1. Soil suspended in porous cup

3. Changes in gas content of chamber are detected

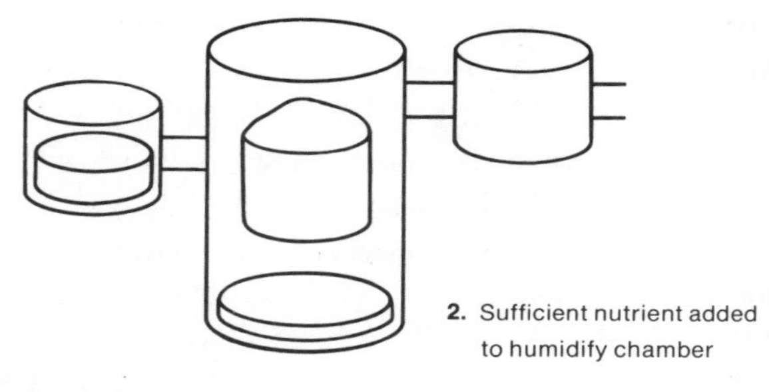

HUMID MODE

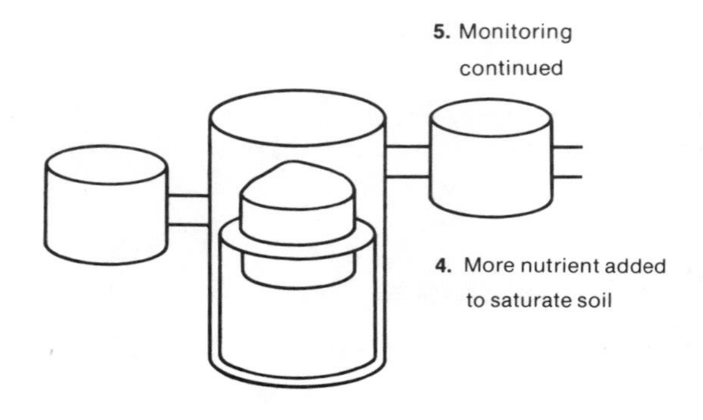

WET MODE

The Gas Exchange Experiment.

periment the nutrient solution consisted of sugars that had been radioactively labeled in the same way as the carbon dioxide in the pyrolitic release test. By including radioactive carbon 14 atoms in the nutrients, this experiment could detect the gases exhaled by any organisms that ate them. This would only work, of course, if the Martian organisms (a) happened to like the sugars used in the test, and (b) were of a type that exhales carbon gases, such as carbon dioxide or carbon monoxide. While these characteristics apply to most terrestrial microbes, there was no way of knowing if they would apply to their Martian counterparts. However, a massive outpouring of radioactive gas was observed—a radiation count of 10,000, as against a background level of 400.[18]

According to Dr. Gilbert Levin, designer of the experiment, "The response that we get, in amplitude and shape, is consistent with the kinds of response we are used to seeing in terrestrial soil." [19] In other words, it was just the response that would be expected from living organisms. In fact, according to Levin, "it suggests some very robust forms of life in the soil." [20]

The reaction was eliminated by sterilization, once again adding strength to the biological interpretation. As Dr. Klein described it, "First, we got a process that looked either biological or chemical. Then we did the one thing that we felt would discriminate biology from chemistry, and it came out like biology." [21] But, at this point, there was still a slim chance that a chemical reaction could have caused these results. There are a handful of chemicals that could have produced a similar release of carbon dioxide and which could have been destroyed by the high temperatures used for sterilization.

To eliminate this possibility, another test was devised. The sample was sterilized at a lower temperature: 122 degrees instead of 300. The idea was that any organisms in the soil would be harmed by this heat, and therefore would not produce the same strong results. On the other hand, most of

LABELED RELEASE

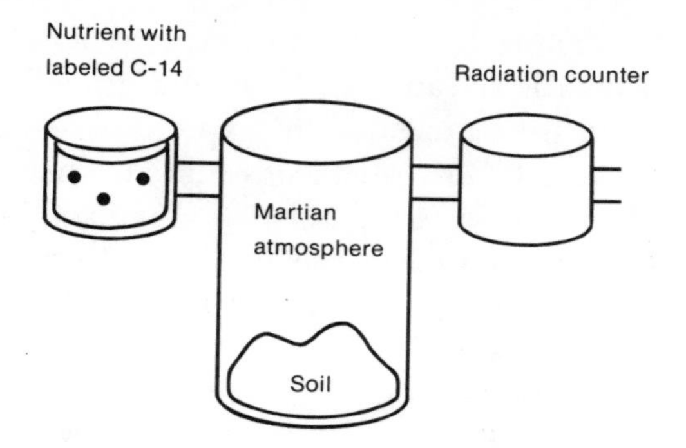

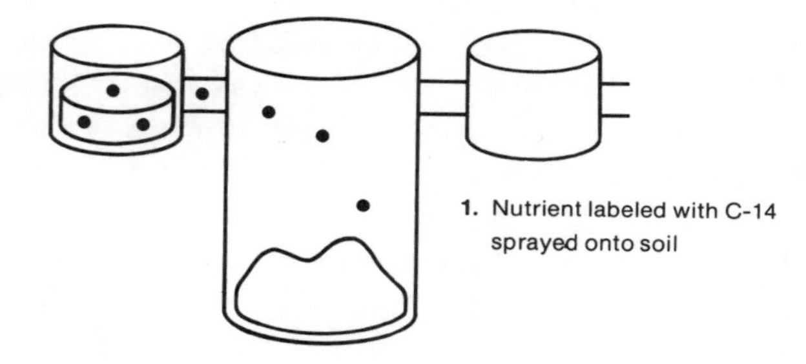

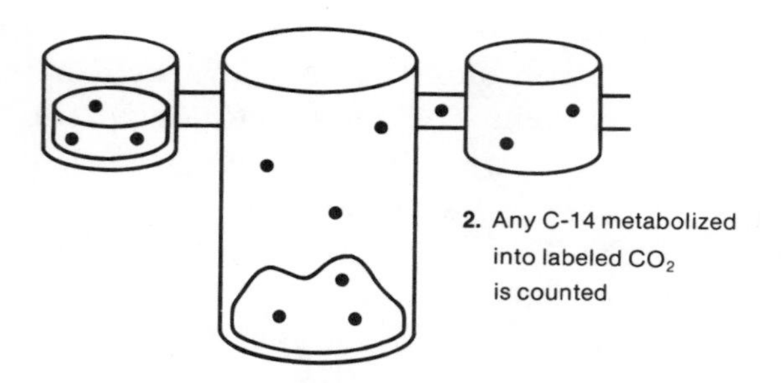

The Labeled Release Experiment.

the chemicals that might have been responsible for the reaction would not have been affected at all by this lower temperature.

The result was one of the great surprises of the mission. It showed a level of activity drastically reduced from the original level, thereby ruling out most of the possible chemical explanations. But it was also very different from the expected biological result. What happened was that the carbon dioxide was alternately given off and reabsorbed by the soil, in a regular one-day cycle.[22] This is similar to the behavior of certain terrestrial organisms, which breathe in carbon dioxide during the day and then exhale it during the night.

The surprising thing about this result is that it was so different from the original reaction. If an organism which showed this pattern of behavior were present in the soil, then a similar reaction should have been seen in the initial run.

One interesting possibility is that there may have been more than one kind of organism in the soil producing the original reaction; perhaps two or more organisms existing in a symbiotic relationship. It could be that one of these species was destroyed by heating, but the other was not, and thus gave rise to a very different result than had been seen when the organisms were acting together.

Some of the scientists think that the labeled release test results could have been produced by the same reactive chemicals (peroxides, ozonides, or superoxides) that had been postulated to account for the gas exchange test results. But others, like Dr. Robert Jastrow, do not think that is possible: in the results from the second landing site, the gas exchange test produced only one-tenth of the original response, while the results of the labeled release test were the same as those at the first site. If both of these reactions were caused by the same set of reactive compounds, as had been theorized, they should have changed in the same way. According to Jastrow, in his book "Until the Sun Dies." "This result seems to indi-

EARTH

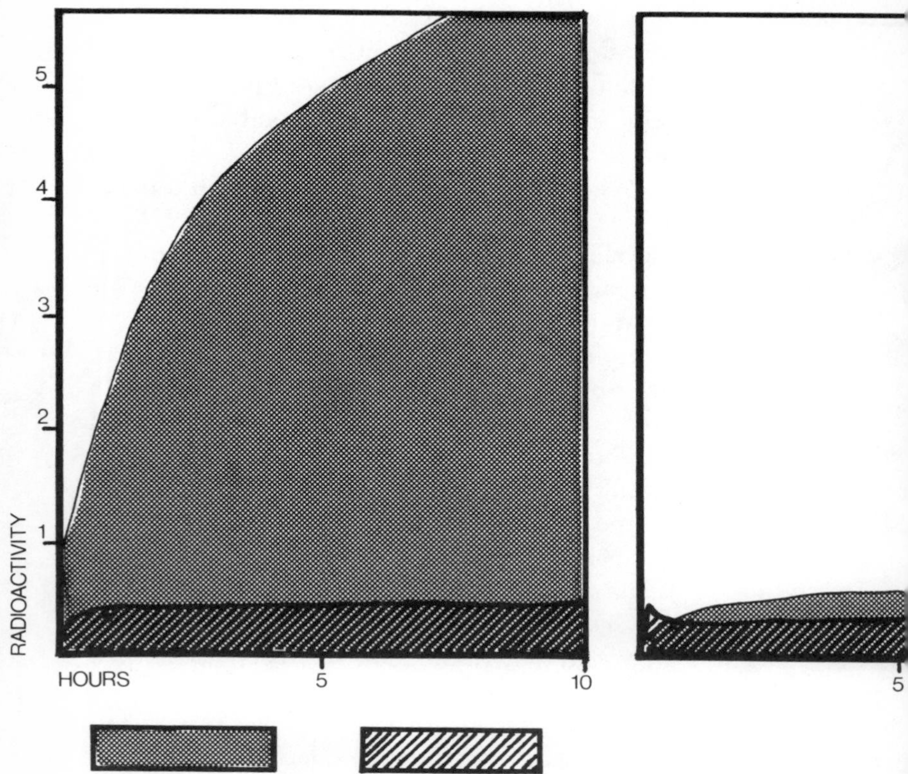
MOON
5
4
3
2
1
RADIOACTIVITY
HOURS
5
10
5
ACTIVE

CONTROL

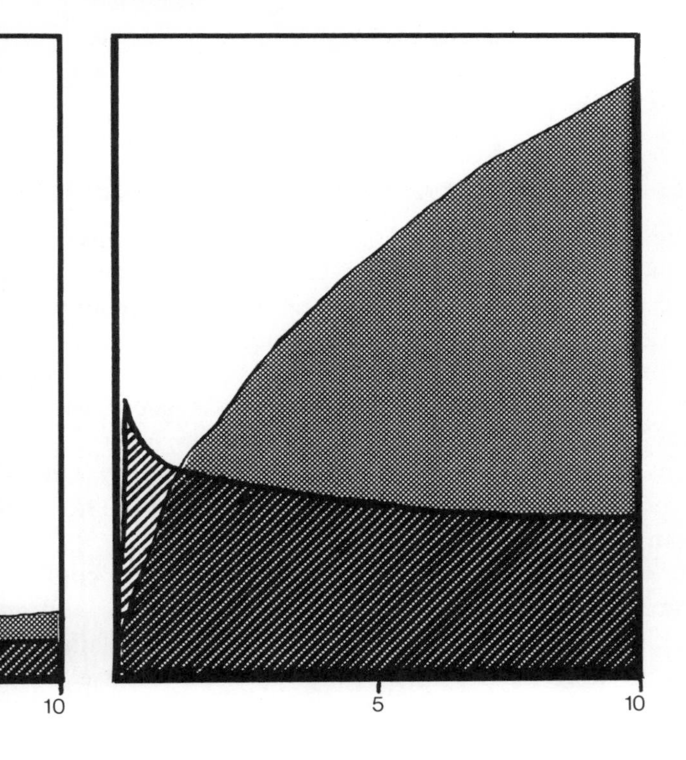

Labeled Release Experiment results. The results of the labeled release test are shown as a graph of radioactivity (counts per second) versus time. Results of the first Viking I test on Mars correspond closely to the results from Earth soil known to contain living organisms. The most significant feature in this test is the difference between the results from active runs, where life should flourish, and control runs, in which the soil was first heat-sterilized to kill off any living organisms. In the results from lifeless lunar soil, unlike the Earth and Mars soils, there is virtually no difference between active and control runs.

cate that chemical reactions involving peroxide compounds cannot be the source of the lifelike signals obtained in the microbe (labeled release) test. With the chemical theory for the test eliminated, a biological process is the most straightforward explanation remaining." [23]

Organic Chemistry: Where Are the Bodies?

The one stumbling block to a biological interpretation has been the results from the GCMS (gas chromatograph mass spectrometer) which analyzed the soil for large amounts of organic chemicals (the decayed remains of billions of dead organisms) and which failed to find any. According to Viking chief scientist Gerald Soffen, "All the signs suggest that life exists on Mars, but we can't find any bodies!" [24]

This is the center of the controversy over Martian life—the supposed conflict between the biology experiments and the organic chemistry analysis.

But the fact is that these results are not really in conflict. These are different kinds of experiments, and they are telling us different things about Martian conditions. As Klaus Biemann, who designed the GCMS, has repeatedly stressed, this device was never intended as a life detection experiment. The results from the GCMS do not contradict the life tests. On the contrary, there are a number of perfectly reasonable ways to account for this result without denying the presence of life.

1. The GCMS is not looking for living organisms, and could not detect them even if they were present. The amount of living matter in a rich sample of garden soil on Earth (about ten million microbes per gram of soil) would be much too small for this instrument to detect. However, on the Earth, the dead remains of organisms that live in the soil have accumulated unchanged for billions of years, and this mass of carcasses is detectable.

So, if conditions on Mars were just as stable as they are on Earth, we might have expected to find some organic chemicals. But Mars is *not* like the Earth.

The Martian surface is constantly bombarded by a very heavy dose of ultraviolet radiation, which is known to quickly decompose organic molecules, and would thus prevent them from accumulating.

2. The biology experiments that gave the positive results "are a thousand times more sensitive than the organic chemistry experiments," according to Dr. Carl Sagan. Therefore, "there could well be organic chemistry there to explain the results of the biology experiments, but it may be below the sensitivity of the organic experiment."

3. There are a number of long organic molecules which would be decomposed by the heat of the GCMS itself, and would therefore never be detected. On Earth, the only molecules fitting this description are synthetic, but, says Dr. Levin, "they might constitute the backbone of life elsewhere."

4. These valuable molecules may simply be eaten up by living creatures in the soil. Just as vultures abound in Earth's desert regions, we might expect to find scavengers in any area where food is very scarce. In the harsh, arid Martian environment, an organism would have to be adept at searching out and quickly devouring every available scrap of organic material. This kind of efficient recycling would cut down on the buildup of waste products, and perhaps leave too little for the GCMS to detect, even if the soil were teeming with a large variety of active, healthy creatures.

5. Life may exist on Mars, but only in certain pockets, or "microenvironments," as Carl Sagan and Joshua Lederberg have suggested. In other words, Mars may be a vast desert with only a few isolated oases, where life can thrive under special conditions in a small area, for example, in a warm pool created by geothermal heating. In that case, a small number of seeds or spores might be transported all over

the planet by the winds, but would only grow and reproduce in these few select spots. Thus, Viking would detect a few spores, but no accumulation of organics.

This conclusion has strong support from the fact that of all the Earth soils that have been tested by the pyrolitic release experiment, the one that comes closest to reproducing the Martian results is Antarctic soil, which includes spores transported by the winds from more fertile areas, but no active life.

Another theory is that Martian organisms might have developed thick shells to protect themselves from the deadly ultraviolet radiation, and that these shells might prevent the organic remains from reaching the GCMS and being detected.

In short, there is no real conflict between the biology test results and the organic chemistry analysis. These tests simply gave different kinds of information about Martian conditions, and there are a variety of explanations that can accomodate all of the experimental results. The life-detection tests gave the kind of results that might be expected from living organisms, and the organic chemistry test does not contradict those results; it simply places some additional constraints on, and gives us additional information about, the possible nature of Martian biology.

Each of the biology experiments has shown evidence of the biological activities that it was designed to detect. These reactions show many of the characteristics associated with living things (i.e. diurnal cycles, death resulting from sterilization, sensitivity to small temperature changes, dependence on sunlight). And no chemical scheme has yet been discovered that can duplicate the observed results, even though considerable time and ingenuity has been invested in the search for such a scheme.

The Viking mission has yielded considerable evidence to support the conclusion that Mars is a biologically active planet. According to Dr. Gilbert Levin, of the biology team, "The accretion of evidence has been more compatible with biology than with chemistry. Each new test result has made it more difficult to come up with a chemical explanation, but each new result has continued to allow for biology." All of the life-seeking tests showed reactions that, says Dr. Levin, "if we had seen them on Earth, we would unhesitatingly have described as biological." All of them have been confirmed by control experiments. All of them have been successfully repeated at both of the landing sites.

We tossed three fishing lines into the unknown pond of Mars, and we have gotten some strong nibbles. But until we can actually pull the unknown creature out of the water and into the light of day to take a close look at it, there will be those who will claim that it is not a fish at all, that we have just managed to snag the line on an old boot. Arguments are sure to continue, but the question will probably not be answered, once and for all, until the first team of astronauts arrives on Mars for a direct, extended examination.

Their definitive answer is not likely to reach us for at least twenty years. In the meantime, to paraphrase the old adage, if it looks like a fish, swims like a fish, and tastes like a fish, maybe it *is* a fish. Whatever it is we have been examining on Mars behaves, in every way we know of, just like a living organism. Maybe it really is a living organism.

We cannot be quite 100 percent sure, but it certainly looks as though there is life on Mars. So far, the evidence we have in hand only points to the existence of the simplest and most primitive kind of life, microbes. But we may have been examining bacteria in the midst of a rich flora and fauna. We have not eliminated any possibilities about the extent or the level of intelligence of Martian life. All we have done is to raise the bottom line: at the very least, it is quite likely that Mars supports microbial life. At the most . . . nobody knows.

The Parade of the Seasons

> *The other planet [than Earth] definitely known to support some kind of life is Mars . . . In the minds of most astronomers there is little doubt at this time that the green areas of Mars contain a form of vegetation similar to our lichens.*
>
> —*Dr. I.M. Levitt*
> "A Space Traveller's Guide to Mars"

FOR MORE than two billion years after life began on Earth, nothing more advanced than microbes lived on this planet. Then, beginning with the first photosynthetic plants, all the diversity of life as we know it today evolved in only half that time: about one billion years. The development of plant life thus seems to mark the turning point from primitive microbial life to a richly developed variety of advanced life forms, and thus, eventually, to intelligent life.

The Viking experiments showed evidence for the existence of microbial life on Mars. That would make it much likelier than most people had previously believed that more advanced forms might also have evolved there. If plant life were to be discovered on Mars, it would greatly improve the odds for finding more advanced kinds of organisms. So it is now more crucial than ever to see if there is any sign of vegetation on Mars.

For over a century, many scientists considered the seasonal waxing and waning of dark markings on the face of Mars to be evidence of the existence of some kind of plant life on that planet. These dark markings, which have been known since the seventeenth-century observations of the Danish astronomer Christian Huygens, cover more than a third of the surface of Mars. But their darkness, color, and extent are highly variable, and seem to change with the seasons of the Martian year.

It was these seasonal variations that led observers to speculate that the growth and decay of plant life might be involved. This idea was first proposed in 1860 by astronomer A. Liais, but was not widely accepted until the theory was popularized by Percival Lowell around the turn of the century. For several decades, this was accepted as the standard explanation of Mars' changing dark patches, and the belief in Martian vegetation was nearly universal.

In recent years—beginning with the bleak view of Mars presented by the Mariner 4 photographs, which convinced many people that life was impossible there—this view seems to have fallen from favor. But, in many ways, the evidence for Martian plants is now stronger than ever.

No one questions the fact that the dark markings change, and that the changes tend to involve growth during the spring and summer, and decline during the fall and winter. This pattern, in itself, certainly is reminiscent of the seasonal growth and decline of vegetation in the temperate zones of the Earth.

On Earth, the wave of new growth spreads from the tropics toward the polar regions, as the sun's warmth reaches further north with the lengthening days. But on Mars, the trend is in reverse: dark areas first increase near the pole, and then the darkening spreads gradually toward the equator. This has long been known as the "wave of darkening." Its opposite direction from that seen on Earth has been explained by the fact that here, warmth is the limiting factor for plant growth, whereas on Mars the limiting factor may be the availability of water.

As the Martian polar caps melt in the approaching summer, the air becomes saturated with water vapor—starting at the fringe of the melting ice cap, gradually being carried outward by the wind, and ultimately reaching all the way to the equator. These humid winds, blowing across fields of dormant vegetation, might be just what is needed to bring them back to active life.

Recently, there has been some question as to the reality of the wave of darkening. In fact, it is clear from recent observations that it is not a "wave" in quite as strict a sense as was once thought. There is no continuous, constant spread of darkness reaching from pole to equator. However, precise photometric studies of hundreds of Mars photographs have shown a definite, statistically significant trend in which the dark areas near the pole are most active in the spring, and those near the equator are most active in the summer.[1] So, at least in its broad outlines, the wave of darkening appears to have been confirmed.

One great surprise from the Mariner 9 mission was that the dark markings, which are so clearly observed from the Earth, do not seem to correspond to any particular features of the Martian geography seen at close range.

They are neither high nor low, but are distributed quite randomly among some of the highest and lowest parts of the Martian terrain. Some dark areas are plateaus, while others are steep slopes; some are mountainous, while others are in deep basins. Their nature is thus a great deal more complex than had been imagined, and they may, in fact, not all be caused by the same phenomenon. Some dark features may turn out to be simply areas of darker volcanic rock; others may be areas created by wind erosional processes; and still others may, indeed, be vegetation.

Upon close inspection (by means of Mariner 9 and Viking

orbiter high-resolution photographs), all of the dark markings turn out to be fragmentary. They are of two types, which have been described as "splotches" and "streaks."

The splotches are irregular patches of dark material, often with ragged edges, arranged in seemingly random ways across the landscape. They do not appear to be noticeably affected by the local topography: they are not related to any of the hills, craters, and fissures in the land.

The streaks, on the other hand, are almost always associated with ridges or craters. For the most part, they form wide "tails" that seem to emanate from craters, giving an appearance that resembles a comet or a flower petal. There is no doubt that these streaks are, in some way, controlled by wind patterns.

There are also light streaks, which are very different in form from their dark counterparts. They are much longer and narrower, and also much more durable. Because of the great differences between the two, it is quite likely that the light streaks are caused by an entirely different process than the dark ones, although clearly the wind is involved in some way in both cases.

Many scientists currently believe that all of the different kinds of dark regions are caused entirely by the action of windblown particles of dust and sand.[2] This explanation, however, runs into some serious difficulties.

The Windblown Dust Theory

There are two basic mechanisms that are believed to be at work that may be causing these dark regions: wind deposition of sand of a different color over the surface material (i.e., dark sand deposited on a light surface, or light sand on a dark surface); or scouring, in which a thin layer of sand is blown away from a surface material, such as bedrock, of a different color.

Light sand deposited by the wind may very well explain

the light streaks. But the transportation of dark material by the wind is extremely unlikely. In all the centuries of telecopic observations of Mars, and in the more than two years of close-up observations by Mariner 9 and the Viking orbiters, hundreds of light-colored dust clouds have been seen, including the globe-circling dust storms that occur nearly every Martian spring. There is no question that light material is frequently stirred up and carried around by the winds. But not once, in all of these observations, has there *ever* been a report of a cloud of dark material.

And yet, it is only the dark areas that are seen to undergo major, rapid changes on the Martian surface. In order to account for these massive and widespread changes, a vast amount of dark material would have to be blown around every year. Surely, if these dark clouds existed, at least one of them would have been seen by now.

One suggestion that has been made by proponents of this theory is that the dark and light materials might really be made of the same stuff, but that a difference in the size of the particles causes the different darkness—the larger grains would give a darker appearance. In that case, the clouds of dust might seem light when viewed from above simply because the smaller, lighter particles were lifted higher by the wind, forming the top surface of the cloud. But this view was refuted by a spectrographic study by Thomas McCord and his colleagues at the Massachusetts Institute of Technology,[3] which established that, while the dust clouds are made of the same material as the light surface areas, the dark surface areas have an entirely different chemical composition.

The fact that dark material has never been seen blowing around thus effectively rules that out as the mechanism for the growth of the dark areas. What about the alternative explanation, that winds scour light material away from a dark base? Here, too, there are problems. First of all, it is very difficult to explain why dark areas appear in the same place each year, and expand in fairly predictable ways. This would make sense if the dark areas were all mountainous or steeply sloping regions. Then, the wind might regularly sweep the slopes clear of an overlying blanket of dust, which would simply be blown away down the slopes. But how to account for the many dark areas that are perfectly flat? For example, Mare Acidalium is one of the largest and darkest of the areas, and yet it is a smooth, low-lying plain. In fact, it is one of the lowest points on the Martian surface, and is part of the basin of the ancient northern ocean. Because of its depth, one might expect dust to accumulate there, rather than being scoured away. No satisfactory explanation has been advanced to account for the darkness of this region on the basis of wind scouring to reveal a dark base material.

The dark streak markings are also very difficult to explain with this theory. It is certainly possible that the wind could scour light dust away from the ramparts of a crater. But the long tails streaming away from the downwind side of the crater just do not make sense. It has been well established by wind-tunnel tests that dust should be *deposited* downwind from the crater. This is, in fact, what seems to happen in the case of the light-colored tails. But scouring of light sand away from dark rock in the lee of the crater contradicts the known flow patterns of the wind past a crater-shaped obstacle.

The only way to account for the dark streaks in terms of wind-transported dust would be if the crater itself were a *source* of dark dust. But this does not make sense either, because the craters are not volcanic, but rather the result of meteor impacts; so there is no deep, long-lasting source of material that could emanate from the crater, as would be the case with a volcanic vent. And if there were simply a static source of dark material lying inside the crater, it would have been used up long ago: this process must have been going on for millions of years.

There is also a major problem in accounting non-biologically for the survival of dark areas after a dust storm. If the dark markings were just surface material, they should disappear after being covered by a large cloud of light powder for a period of several weeks, as happens during Mars' frequent global dust storms. But that does not seem to happen: dark markings almost always survive the dust storms. For example, a new dark streak was seen in 1973 near the Martian equator, extending for about 3,000 miles. Shortly after it was first seen, Mars was covered by a planetwide dust storm that lasted almost a month. But when the dust subsided, there was the new marking, just as long and as dark as before.[4] It is easy enough to picture plants that could grow right through a freshly deposited layer of sand, or larger plants forming a kind of forest in which the sand would filter down between them without covering them up, but it is pretty hard to imagine a layer of black sand or rock that could recover its darkness in the same way.

The Moist Breath of Life

Even though windblown dust does not seem to explain the phenomenon, there is no doubt that, in some way, the wind does have an effect. In the case of the dark streak markings, all of the tails always seem to be aligned in one direction, and clearly result from some aeolian process. If it is not dust being blown around, then what is it?

The obvious answer is that moisture in the air is responsible for the growth of plants in the path of the wind. The fact that dark tails tend to emanate from craters makes sense if the craters, which are generally deeper than the surrounding land, may contain a reservoir of water. When the sun caused this water to evaporate, the prevailing winds would carry it downwind from the crater where a wide swath of plants could benefit from the life-giving humid air.

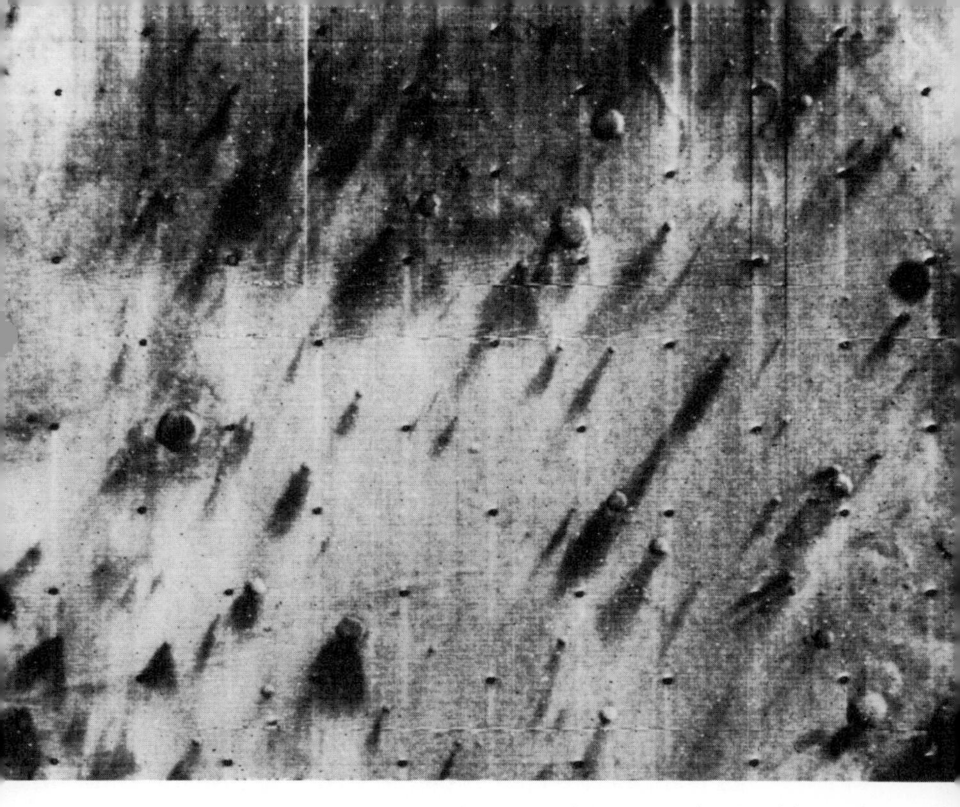

These typical dark "streak" markings are clearly associated with craters, and apparently are formed by some process involving the wind.

The presence of water in the crater basins is not just speculation. In many of the Viking orbiter photographs taken just after dawn, a heavy layer of mist is seen inside crater basins, where it presumably results from the evaporation of ice or frost within the craters by the sun's first rays. If a wind were present, it would blow this mist out of the crater in just the kind of tail pattern that is seen in the dark markings.

This may also explain why some of the most prominent dark features, notably Syrtis Major (the darkest and longest-known of Mars' surface features), are found in areas where the prevailing winds blow up from the adjacent lowlands—the former ocean floor, where the ice is most abundant—across a gradually sloping plain. These winds may pick up moisture in the basins, which may then be deposited as they rise over the long, gentle slope. The irregular streaky markings of Syrtis Major would then make sense—their appearance has been compared by the well-known astronomer V. A. Firsoff to the pattern of vegetation seen in Earth satellite photographs of the flanks of the Andes, where moisture laden winds blow up from the shore of the Pacific Ocean.[5] The similar circumstances may, not surprisingly, have created similar results.

Perhaps the most convincing feature of all is the dark collar that forms each spring around the melting polar ice cap, and which follows the edge of the ice as it recedes through the spring and summer. This dark ring has been known since the turn of the century, but the question of its existence had been controversial until the Mariner 9 photographs confirmed it.

This is the strongest single item in the case for Martian vegetation, since this ring around the pole is the single likeliest place for life to flourish on Mars. Because of the rapid melting of the ice, the area right next to the receding mass of ice has a higher moisture content than any other place on Mars. If water really is the limiting factor for the growth of vegetation on Mars, and there's every reason to believe that it would be, then this polar collar should be the first place to look for signs of Martian plants. The dramatic dark ring that we find there is exactly what we would expect, and it is very hard to account for any other way.

The only reasonable alternative is the scouring away of light surface material by the wind. Apart from the objections to this theory that have already been discussed, there is also every indication in this case that there is no dark base for the scouring to reveal.

The region around the North pole, extending down to

about 30 degrees North latitude, is covered by a thick blanket of sediment, which has been estimated to extend to a depth of at least a mile.[6] This sedimentary material is light-colored. There is no way that seasonal winds could scour away a mile of sediment in a few weeks' time. Once again, we seem to be left with plant life as the most reasonable explanation.

One objection that has been raised to a biological interpretation of the dark markings is the apparent absence of chlorophyll in these areas. Chlorophyll is basic to the metabolism of terrestrial plant life. Does its absence in the Martian dark regions rule out the possibility of plant life?

Well, it certainly rules out familiar, terrestrial types of vegetation, but we were not really expecting that anyway. Martian life would undoubtedly differ greatly from terrestrial life. Even so, the fact is that of all the plant life on Earth, the kind that comes closest to being equipped for survival under Martian conditions is the lichen. Lichens are hardy, able to withstand extremes of cold and dessication, and they survive in areas of Antarctica and the Himalayas that come as close as any place on Earth to approximating the surface of Mars. And lichens, as it happens, have very little chlorophyll, not enough of it to be detected by spectroscopic methods.

Furthermore, even on Earth a few alternative photosynthetic molecules have been found, including a purple one used by some microbes. It is now generally believed that there may be dozens, or even hundreds, of possible chemical alternatives to chlorophyll, and that it was just a random event in the early course of evolution that led to chlorophyll's near universality in the Earth's plant life.

Another objection to the notion of Martian vegetation is the absence of oxygen in its atmosphere. All plants on Earth, even lichens, give off oxygen as a natural part of their metabolic processes. If large areas of Mars are covered by thriving plant life, why is there no buildup of oxygen in the air?

Once again, we must remember that Martian life could

not be expected to duplicate earthly life forms. But we can, perhaps, anticipate some of the differences. It was discovered in the experiments with terrestrial plants under simulated Martian conditions (the "Mars jars") that whenever oxygen was present in the air, it severely hindered the ability of plants to survive the cold extremes of Martian temperature. But when there was no oxygen at all in the jars, the plants could tolerate the cold much more easily.[7] So if there is plant life on Mars, it would not be surprising to find that it had evolved with the capacity to retain the oxygen produced by its photosynthesis within its own tissues, instead of letting it out into the atmosphere. And it turns out that, under conditions of extreme cold, even some terrestrial plants do just that.

If Mars is, in fact, the abode of abundant plant life, what are the prospects for finding animal life there? On Earth, plants sustain animals by producing the oxygen they breathe and the carbohydrates they eat. Might Martian plants also support animal life? The chief objection, again, is the absence of oxygen in the Martian air. No earthly animal could survive on Mars today for that reason. But it is certainly possible to imagine ways that animal life could exist without an oxygen atmosphere.

If it is true that Martian plants store, within their tissues, all the oxygen they produce, then an animal might be able to obtain all the oxygen it needed directly from the food it ate. Instead of lungs, it would have an oxygen-exchange system built into its digestive tract. This might not be as efficient a mechanism as our breathing, and Martian animals might therefore be somewhat sluggish by our standards, but it is certainly a possible basis for animal life. We must conclude that, while there is not yet any evidence for animal life on Mars, the possibility of its existence cannot be ruled out.

And at the very least, there is some evidence for a thriving vegetation on Mars today, left over from its halcyon days of warmth and flowing water.

7

Is There Life on Earth?

Let us suppose that a spacecraft is soon to be landed upon Venus or Mars; what more fascinating question than to find out whether our neighboring planets are, or at some earlier period have been, inhabited by intelligent beings capable of projective [tool making] activity? In order to detect such present or past activity we would have to search for and be able to recognize its products, *however unlike the fruit of human industry they might be. Wholly ignorant of the nature of such beings and of the projects they might have conceived, [we] would have to utilize only very general criteria, solely based upon the examined objects' structure and form and without any reference to their eventual function.*

—*Jacques Monod*
Chance and Necessity

MANKIND DOMINATES the Earth; we have tamed the land and harnessed the forces of nature. Everywhere we look around us, the face of our planet has been altered by our influence: from superhighways to strip mines, from irrigation canals to urban sprawl, there is hardly a place left on Earth that does not show some evidence of our handiwork.

We might reasonably conclude that if alien beings ever decided to fly a reconnaissance mission through our skies, it would not take them long to figure out that the Earth is not only inhabited, but that it is inhabited by intelligent beings.

It turns out that it's not that easy. Surprisingly enough, it is quite difficult to find any evidence, from the height of an orbiting spacecraft, for any kind of terrestrial life at all. In order to detect signs of our civilization, those alien beings would need cameras that were much more powerful (that is, cameras with much better resolution) than those that are photographing the Martian surface from the Viking spacecraft.

In other words, if Mars currently had a civilization equaling ours both in sophistication and extent, nothing that we have done so far in our exploration of that planet would have detected any sign of it.

Of course, that does not mean that there is any reason to suppose that there *is* an advanced civilization on Mars. But since we do know that it's quite likely that there's *some* kind of life there, and since biologists agree that once life begins, the development of intelligence and a technological society are probably inevitable, given enough time, it therefore behooves us to look at Mars much more closely than we have so far, and to see just what kinds of life might be there.

The Radio Bubble

The effects of technology do manifest themselves in ways other than those that would show up in satellite photos. For instance, we on Earth put out a great quantity of radio emissions covering a broad range of frequencies. Most of this energy stays close to the Earth, where it's intended to be, so that we can pick it up in the form of television programs, top 40 radio, and CB chatter. But, inevitably, a certain fraction of that broadcast energy always ends up going off into space. This spurious radiation would easily be detected by an approaching spacecraft long before it went into orbit around the Earth, and would signal clearly the presence of intelligent beings. In fact, with sufficiently powerful receivers, this unintentional beacon of radio waves might even be detected across interstellar distances. Since we have been broadcasting for over fifty years now, the expanding "bubble" of radio waves through space (moving at the speed of light) is now a sphere with a radius of fifty light years, which is about 293 trillion miles. That means that any star within that sphere, of which there are about 100, could conceivably pick up these radio waves and determine from the onslaught of big band music and old "I Love Lucy" broadcasts that yet another planet in this galaxy has crossed the threshold into advanced technology.

A receiver capable of picking up our broadcasts over such vast distances is beyond anything that has yet been developed on Earth, however. So, if some of our neighboring stars are sending radio waves our way in the same inadvertent manner, we would not yet know about it. Similarly, even though it is much closer, Mars could be putting out a fairly considerable amount of energy in radio emissions without having yet been detected. We just haven't spent any real effort listening for it.

Even if we *were* to make a concerted effort to detect radio emissions from Mars, a negative result would have little significance: fifty years ago, a scan of the Earth at radio wavelengths would have revealed no activity at all; that could hardly be taken as evidence that the Earth was then devoid of intelligent, civilized life. Fifty years from now, the same might be true: more efficient transmission systems, such as the replacement of television broadcasting by worldwide cable systems, might cut the wasted energy currently radiated into space down to nothing. So, if a Martian civilization were just slightly ahead of us or slightly behind us in their technology, the search for radio emissions might be wasted. Because of this slim likelihood of detecting anything, such a search was long ago deemed not to be worth the effort and expense; as a result, we simply don't know whether Mars is emitting any

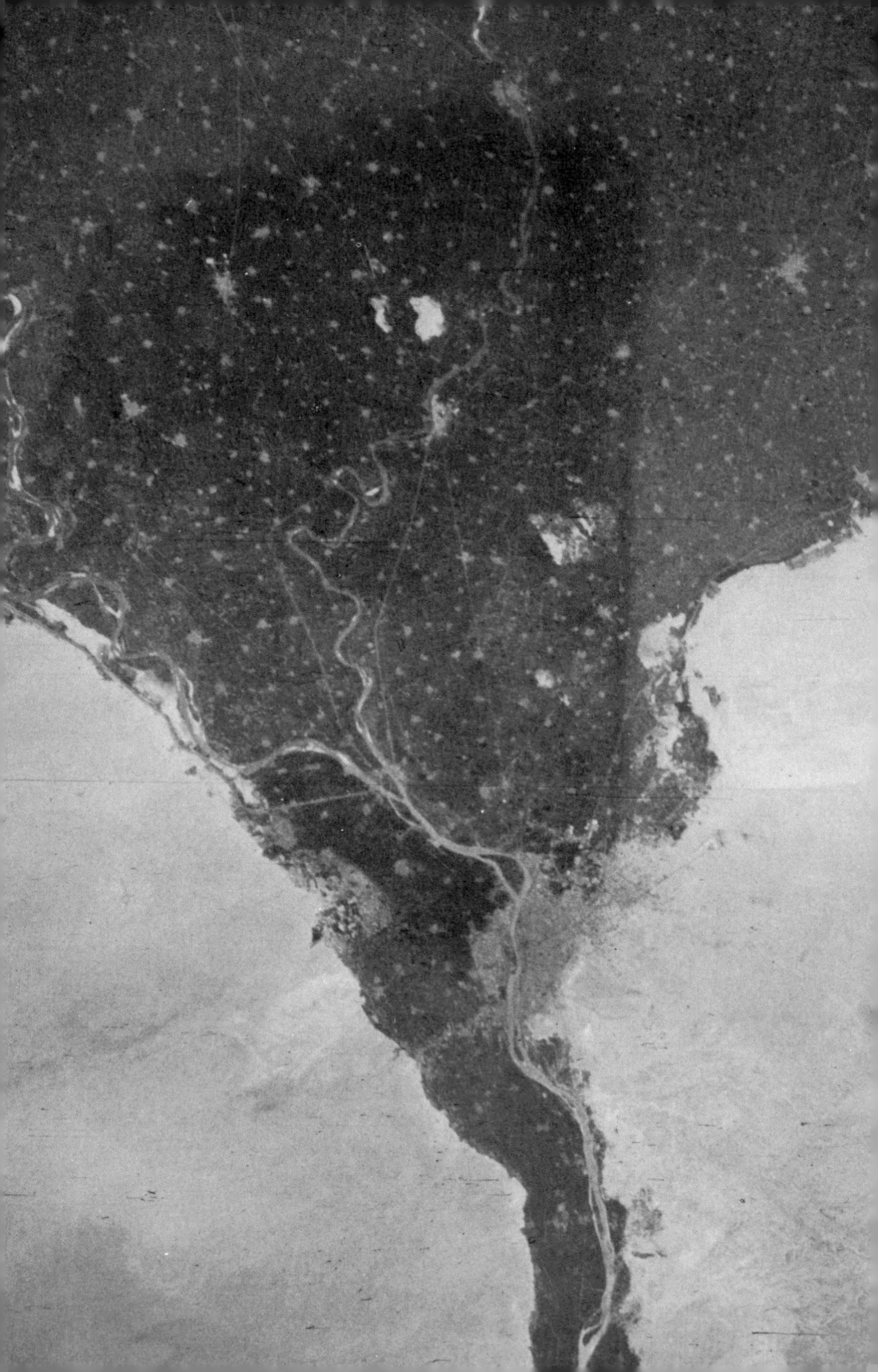

significant radio broadcast energy or not. We just don't know what might be there.

Another sign of technology on Earth is the lighting in our cities. Anyone who has looked from an airplane window at night knows what a brilliant spectacle is produced in the air by the streetlights that we take for granted. During the Gemini program, the citizens of Houston, Texas, all turned out their lights for one minute and then winked them on again as a signal of their good wishes toward the astronauts. The blinking spot of light was clearly visible to them, even though they were hundreds of miles above the city.

Would our orbiting satellites be capable of detecting such a display on Mars? Probably not; first of all, the limits of resolution would probably make it difficult to detect even if the cameras were pointed in the right direction at the right time. Even more important, the cameras are almost certain *not* to be pointed toward any such possible display, for the simple reason that we have not wasted the time and energy of the orbiter cameras by pointing them toward the night side of Mars. The night side is in total darkness, and unless there *were* some flashing city lights, there would not be anything to see. The slim chance of detecting a light could not justify spending all the time it would have taken to produce

Egypt's Nile delta (the large dark triangle of fertile land) was photographed by the Gemini V astronauts with about the same resolution as the best of the Viking photographs of Mars. Cairo, one of the world's largest cities, is visible as a small gray patch where the Nile valley runs into the delta—about a third of the way up from the bottom. The pyramids of Giza, including the Great Pyramid of Cheops, lie just to the left of Cairo but are not visible. This photograph demonstrates the difficulty of detecting evidence of life on Earth by means of orbital photographs with 100 m. resolution.

This Mariner 9 photo shows white streaks that are believed to be caused by windblown dust. The regularity of these features illustrates the difficulty of determining, from photographs alone, whether a given feature results from geological processes or from the actions of living beings.

a selection of photographs that would probably have come out solid black.

Searching for Terrestrial Life

Just how much closer would we have to look in order to find a technological civilization on Mars, if one existed? That question has been answered by a careful examination of thousands of satellite photos of the Earth taken with various degrees of resolution. This research was conducted by Carl Sagan and his associates at Cornell University.

At a resolution of one kilometer, which is equal to the best resolution obtained by all photographs prior to Mariner 9, and also to many of the Mariner 9 photos, the detection of life on Earth is all but impossible. Sagan estimates that about .1 percent of the photographs taken at that resolution show signs of intelligent life (lower forms of life are much harder to detect).[1] But even that may be an optimistic estimate: the Earth photographs which do show signs of life at that resolution are far from being unambiguous.

The best of these photographs, the one that gives the strongest evidence for life on Earth, is a picture of an area in northern Ontario, taken during the winter. It shows a series of parallel lines, which were discovered to be the result of logging activities: the loggers had cut long, mile-wide swaths through the forest, and a subsequent snowfall had increased the contrast between the smooth, fresh-cut swaths and the remaining forest in between, rendering them clearly visible from the air. The result was striking, but even this unusual and dramatic photograph could not be considered absolute proof of life on Earth. Its parallel streaks could conceivably have been interpreted as the result of some unusual cloud formation, for example. Also, some photographs of large-scale dune formations show a similar degree of linearity. If that one photograph were presented as the sum total of evidence for life on Mars, certainly no one would take it seriously.

A further study showed that with a resolution of 100 meters, less than one percent of the photographs showed any evidence of human construction. Furthermore, of the photographs in this one percent, many showed features whose artificial origins could only be established with certainty by correlating them with ground based observations.[2] In other words, based on the photographs alone, it would not have been possible to rule out the possibility of a natural origin for the observed markings.

This is demonstrated by the fact that in this study there was a large number of "false positives," which is to say photographs that showed regular markings that appeared to be artificial, but which turned out to be of natural origin—sand dunes, peninsulas, sand bars, and cloud formations. In fact, the number of false positives was about equal to the number of real positives. The natural markings were very similar to those that turned out to have been man-made. Thus, to an orbiting alien with no direct knowledge of the surface, it would be very difficult to obtain unambiguous evidence for intelligent life on Earth.

Most of the photographs that did show signs of life were of two kinds: roads, and rectilinear patterns of agricultural land. Of these, the photographs of roads are the most ambiguous: although many of them are long and straight, it is easy to imagine a team of Martian scientists looking at these pictures and concluding that they must represent natural fault lines in the Earth's crust. We know, by comparing these photographs with our direct information about the land, that they do show man-made constructions; but it would be impossible to prove that from the photographs alone.

The rectangular fields of large-scale agriculture are the only truly convincing evidence for terrestrial life. These patterns are widespread, and they break the countryside up into a sort of Cubist landscape. The crisp, hard edges of these fields, and their sharp 90 degree corners, cannot by any stretch of the imagination be interpreted as natural formations. If there were widespread agriculture on Mars conducted in the

This Viking I orbiter photograph shows a serie. of parallel grooves (bottom left), resembling contourec farmland. They may be sand dunes, but their regu larity is surprising.

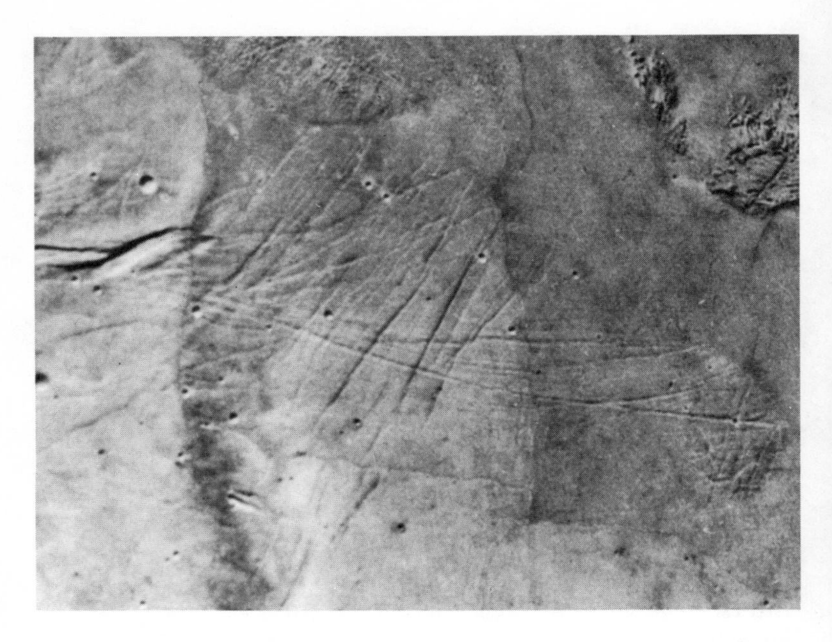

These linear markings appear on many areas of the Martian surface. They may be fault lines in the rock.

earthly manner, with mechanized farming techniques and the large, regular plots of land that mechanization requires, then its effects would be detectable. If we photograph the whole planet at 100 meter resolution (only a small fraction of it has so far been examined that closely), then, if such fields exist, we will find them.

But, if we fail to find these grid patterns, that will certainly not rule out the existence of life, or even of intelligent life and a highly developed civilization.

First of all, there is no compelling necessity for agriculture to be carried out in a grid, though we, at this point in our development, often find it expedient to use such a pattern. For most of our history, agricultural land has not been so visibly divided: in the past, small farms predominated and the great rectangular sprawls of modern farming did not exist; and most farms followed, to a much greater degree than they do now, the natural contours of the land. Such farms would not be detectable from an orbiting satellite.

So it is entirely possible that we might photograph every inch of the Martian surface at the best resolution now attainable, and not discover any sign of life, even if there were presently a Martian civilization comparable to our own in extent and sophistication, but with different methods of agriculture. Even large modern cities could not be unambiguously discerned, and older, less highly structured cities such as Cairo would not show up at all in such pictures.

Since less than one percent of the Earth photographs showed any signs of life, even ambiguous ones, it is significant that only a small fraction of the Martian surface has so far been photographed at high resolution by the Viking orbiters. The fact is that our photographic surveys of Mars, even if extended to the maximum coverage possible from the present mission, will not be very likely to tell us anything definite about the extent or the level of intelligence of possible Martian life. And if it were to turn out that Martian life exists primarily underground, as might be expected because of the harsh surface conditions (especially the ultraviolet radiation) then even a tenfold increase in the resolution of our photographic surveys, perhaps even a hundredfold increase, might still not be enough to reveal it to us. Even highly advanced life may prove to be very elusive.

But even in the areas we have looked at, and even at this low resolution (which could not show anything smaller than a football field), some strange and startling features have been seen. One Viking picture shows a set of curving parallel lines that resemble the furrows of plowed farmland, but on a much larger scale. Perhaps a large terraced hillside would give a similar appearance, but it is hard to imagine what natural process could have caused such regularly spaced lines. Regularity is always suggestive of intelligence, but, as proved

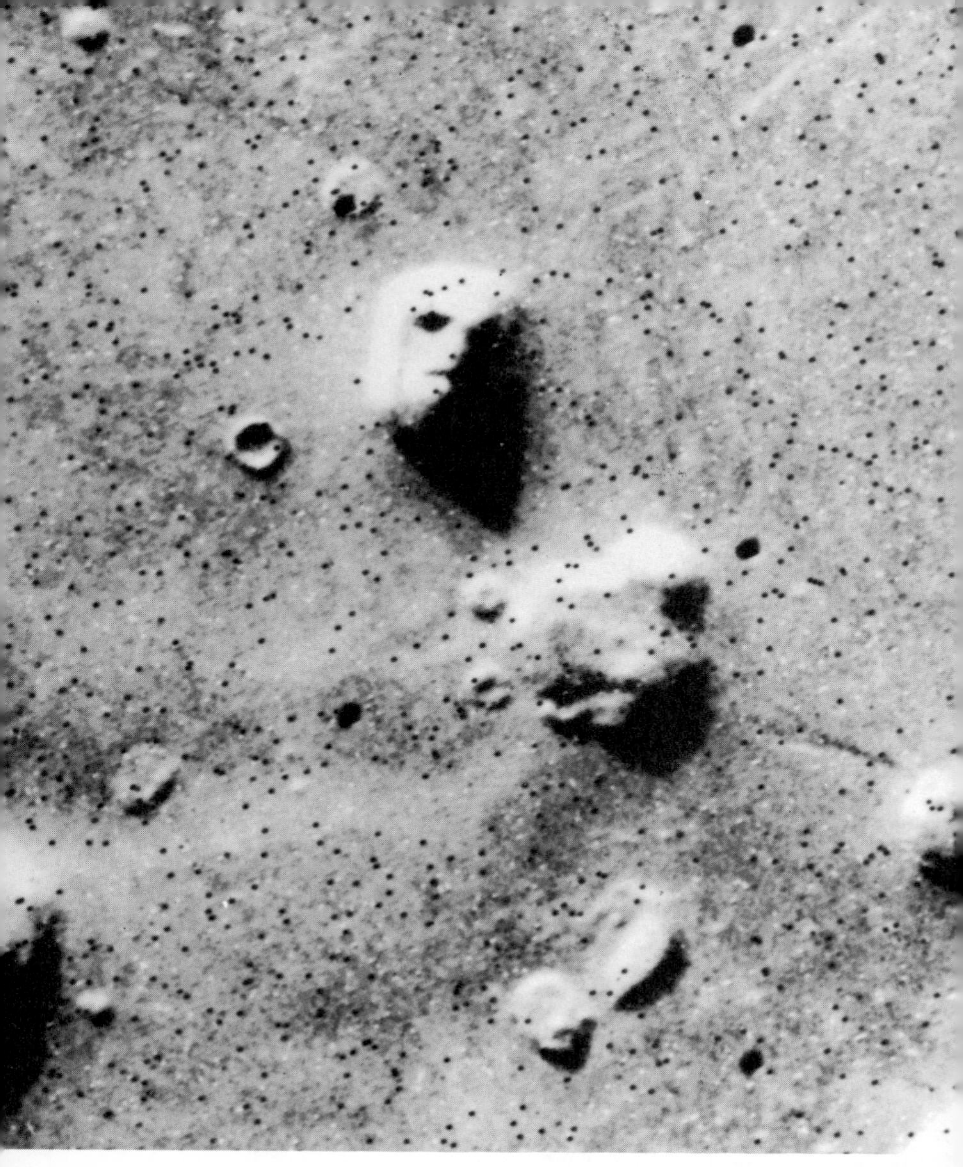

This "human face" is actually a mesa formation, with shadows giving the illusion of human features. Its only significance is as a warning that we must be wary of reading too much into photographs.

by some of the dune formations seen in satellite photos of Earth, it does not in itself prove intelligence.

In photos from many different parts of the Martian surface, there are filamentary markings—thin lines which, sometimes singly and sometimes in large numbers, trace across the land for great distances.[3] In some cases, isolated lines may be found in otherwise featureless terrain. In other areas, there may be many crisscrossing lines. Sometimes these seem to radiate from a central point, sometimes not.

Many of these lines, and perhaps all of them, could simply be fault lines caused by stresses in the rock. But it is highly unusual for such fault lines to radiate outward in all directions, as some of these do. Faults are usually aligned in one direction, representing a weakness in the formation of the rock structure. A radiating pattern is more difficult to explain.

We must always be on guard, however, against reading too much into photographs of a place that we know so little about. A good illustration of this danger is the mesa formation shown in Photograph 20, which resembles a human face. When Gerald Soffen, the Viking chief scientist, first saw this picture coming in from the Viking orbiter, he immediately called up the wire services to announce, with Puckish glee, that he had just received the first actual photograph of a Martian.

Needless to say, the apparently human features seen in that picture were a pure coincidence. A photograph of the same area taken a few hours later, when the sun's angle had changed and altered the play of shadows, showed a very ordinary mesa—an erosional remnant of the highlands of a former epoch, rising from a plain that may once have shimmered with flowing water. The face was just a trick of lighting, a fleeting illusion.

Not all of the photographs can be so easily explained. In

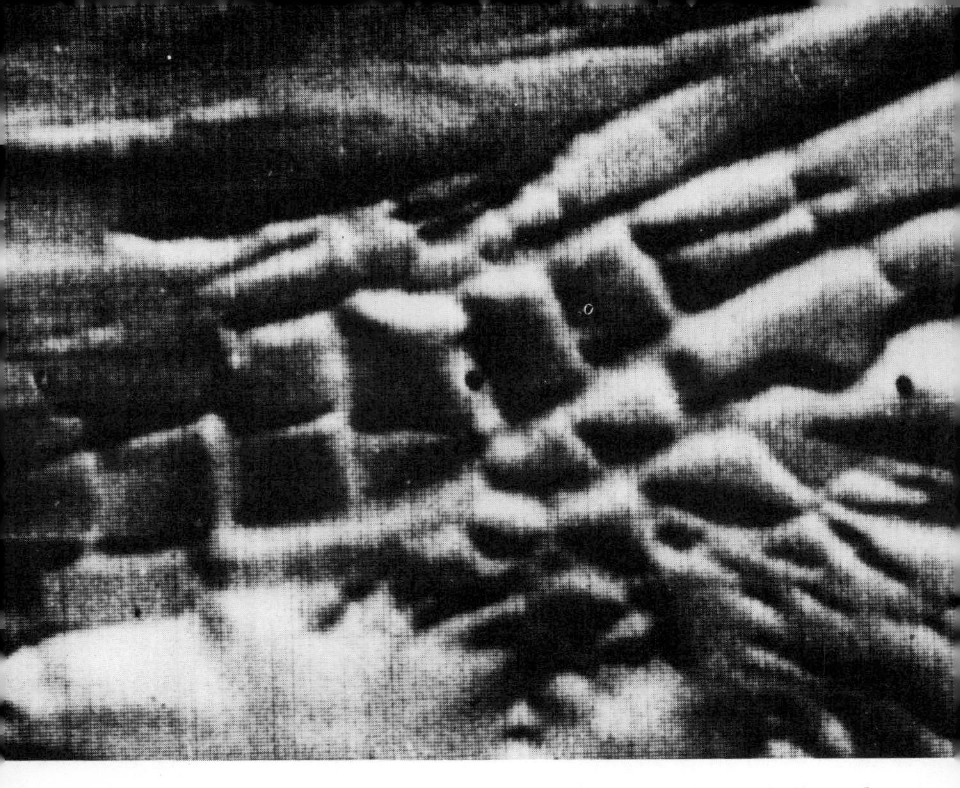

particular, one of the features that has so far defied explanation is a set of amazingly linear ridges near the Martian South pole that are arranged in a kind of grid, forming square and rectangular boxes.

John McCauley, one of the geologists in the Mariner 9 mission, described the feature thus: "The ridges are continuous, show no breaching, and stand out among the surrounding plains and small hills like walls of an ancient ruin. The origin of the reticulate pattern is problematic . . ." [4]

The startling symmetries of this formation were first seen in Mariner 9 photographs, and the scientists involved in that mission referred to the area among themselves as "Inca City." [5] It certainly does resemble an archeological excavation of ancient city walls, but the scientists were not seriously proposing that as an explanation. Most geologists believe that such formations could result from erosional processes in which

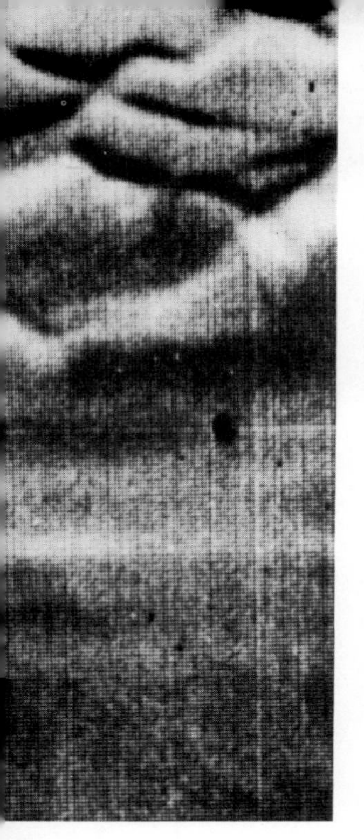

This formation of steep walls and boxed-in areas was referred to by NASA scientists as "Inca City." It may be a geological formation, but its regularity and the straightness of its edges are very hard to explain.

igneous dikes—walls of volcanic material that seeped up through cracks in the rock and then hardened—were exposed because they were harder than the surrounding rock, and therefore were not worn away as rapidly.

But this theory does not explain the rectilinear pattern formed by these walls. A few straight walls would not have been too surprising, and maybe even a right angle or two. But the regularity of the many squared off corners, and the many boxed-in areas of approximately the same size, are difficult to account for. The patterning is reminiscent of the grid formations seen in terrestrial agriculture, and in much of human architecture and engineering, but on a much larger scale: each box in "Inca City" is about three miles across. That makes them about ten times larger than any human construction of similar regularity and symmetry. But no natural formation has been seen on Earth that exhibits such a pattern, even on a smaller scale. Geologists are still baffled by the origins of this patterning.

Such uncertainties are tantalizing. But the most mystifying, and the most suggestive, features yet discovered on Mars were photographed by Mariner 9 on the plateau called Elysium: a group of gigantic pyramids.

The Pyramids of Elysium

Amongst features on Mars which defy positive explanation are three-sided pyramid-like structures in the east central portion of Elysium Quadrangle. The features are seen on the Mariner 9 photographs, (and) cast sharp triangular shadows . . . these tetrahedron pyramids [have a] base width (of) approximately 3 km.

—'Pyramids' of Mars
Spaceflight, May, 1976

IN THE MIDST of the great ocean that once covered most of Mars' Northern hemisphere there was a single continent, about the size of Australia. Rising in a verdant splendor from the depths of the globe-circling sea, this nearly circular landmass may, in those halcyon days of warmth, have richly deserved the name that was given to it many millennia later: Elysium.

The plateau of Elysium is a volcanic bulge in the crust of Mars, similar to the much larger Tharsis bulge a thousand miles to the east, but much older than that still active region. Along one edge of Elysium, following the coastline of the ancient continent, is one of the most active regions of seasonal color change: a dark marking known as Trivium Charontis. This area of changing darkness, whose waxing and

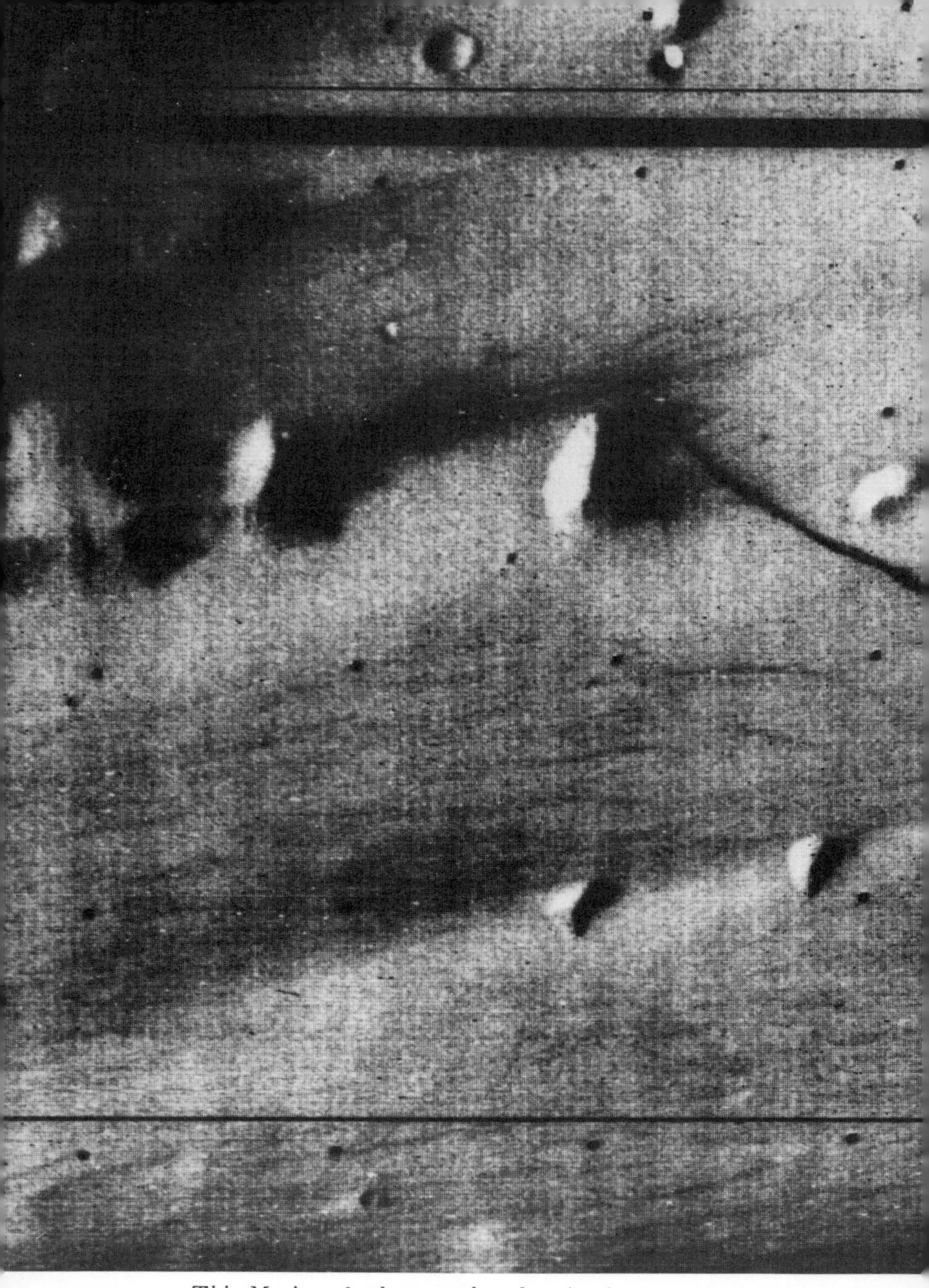

This Mariner 9 photograph, taken in the area of Elysium, reveals the most astonishing and inexplicable features ever seen on Mars: three-sided pyramids, the largest of which are two miles across and

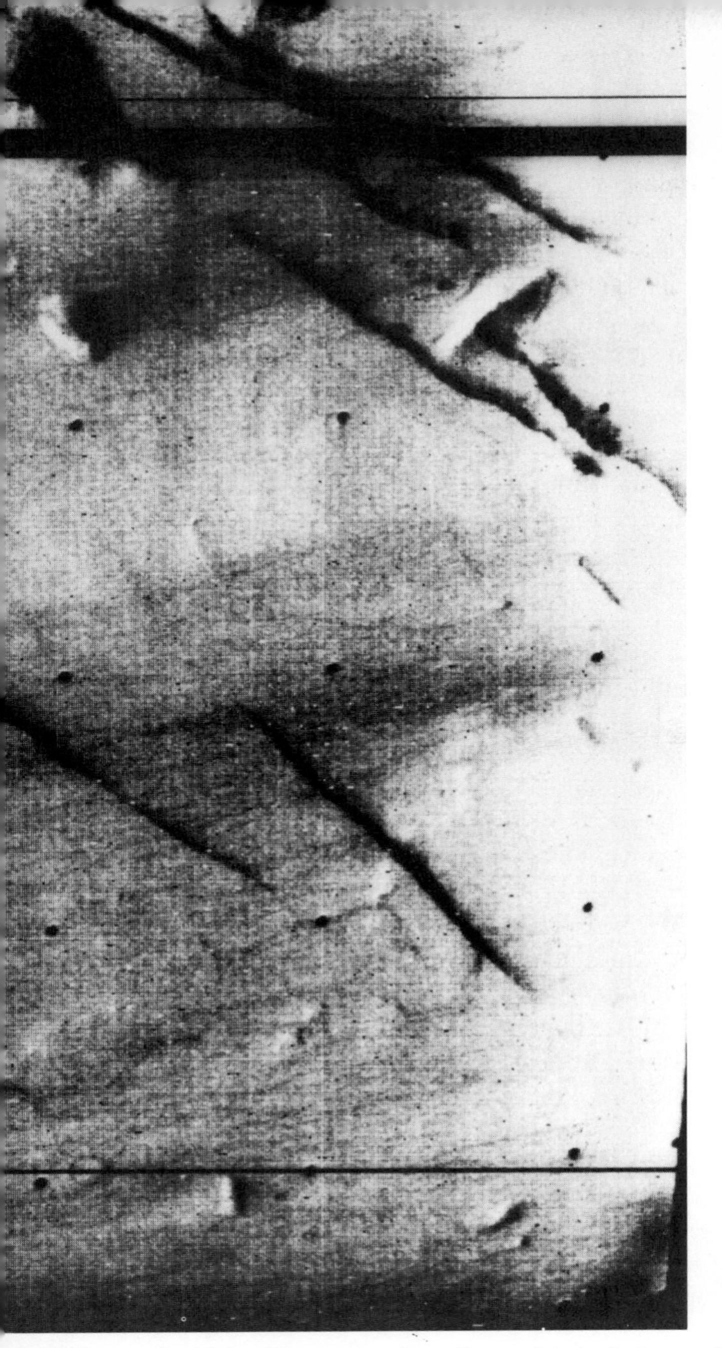

half a mile high. There are also other related shapes such as the rectangular pyramid at top, right of center. The horizontal black line is an artifact of the camera transmission.

waning have been observed for a century, may be one of the most fertile areas of plant life on Mars today.

Rising high above the plateau, near the coastline in the Trivium Charontis region of Elysium, stand the most amazing, inexplicable, and mind-boggling structures yet seen on the surface of Mars: a group of immense, three-sided pyramids that dwarf any man-made structure on Earth, and whose origins have not yet been plausibly explained. The largest of these tetrahedral pyramids are more than half a mile high, more than twice the height of Manhattan's World Trade Center towers. If one of these pyramids were transported to Manhattan, its base would extend all the way across the island, from the East River to the Hudson. It would cover an area of more than 120 city blocks.

Like the great Egyptian pyramids of Giza, the pyramids of Elysium are scattered with no apparent pattern across a smooth sandy plain, and occur in a variety of different sizes but identical shapes and proportions. They are surrounded by other related shapes: there is no Sphinx, but there is a four-sided rectangular pyramid and a variety of irregular polygonal forms.

There is no comparison in scale: the Great Pyramid of Cheops—the largest pyramid on Earth—is about 480 feet high and 760 feet across. The great pyramids of Elysium are almost ten times higher and broader, and occupy a total volume that is nearly 1,000 times larger.

How did these colossal structures get there? If they had been found on an aerial photograph of some unexplored part of the Earth, we would assume that they had been built by some ancient civilization, and teams of archeologists would immediately rush to the spot. But they are not on Earth, they are on Mars. Could they have been created by the natural play of geological forces?

In an article titled "Pyramidal Structures on Mars" (which appeared in *Icarus,* the leading scientific journal devoted to planetary research), Mack Gipson, Jr., and Victor Ablordeppey present four kinds of geological processes that they believe might have been capable, singly or in some combination, of producing such structures.

The Geological Theories

The first theory, and the one that has been most widely accepted as the probable explanation for these pyramids, is that they result from "wind faceting of volcanic cones, lava flow ridges or outcrops by prevailing or storm winds." [1] In other words, the sandblasting effect of windblown dust may have worn down the sides of some natural feature, such as a volcanic cone, so that over a long period of erosion a flat face was produced. Then, presumably as the result of a shift in the direction of the prevailing winds, another facet was carved, and finally another shift in wind direction would complete the three-sided structure. Alternatively, the wind may have flowed over the surface of the object in some intricate pattern, sculpting all three faces at the same time, perhaps as the result of some unusual shape of the original structure.

The example usually cited to support this theory is that of the wind-faceted mountains of the Peruvian Desert. These mountains do, indeed, show some characteristics that are similar to the Martian pyramids: some fairly flat faces, and even a few shapes that are approximately tetrahedral. But all of these structures are part of long connected chains, which exhibit clear patterns of flow along their length. The Martian pyramids, on the other hand, are separate free-standing structures. Furthermore, the Peruvian mountains are much less regular in shape, and much smaller than their supposed Martian counterparts. It seems unlikely that such precise symmetries could have been produced in this way, especially on such a large scale.

But the decisive argument against this theory is also the

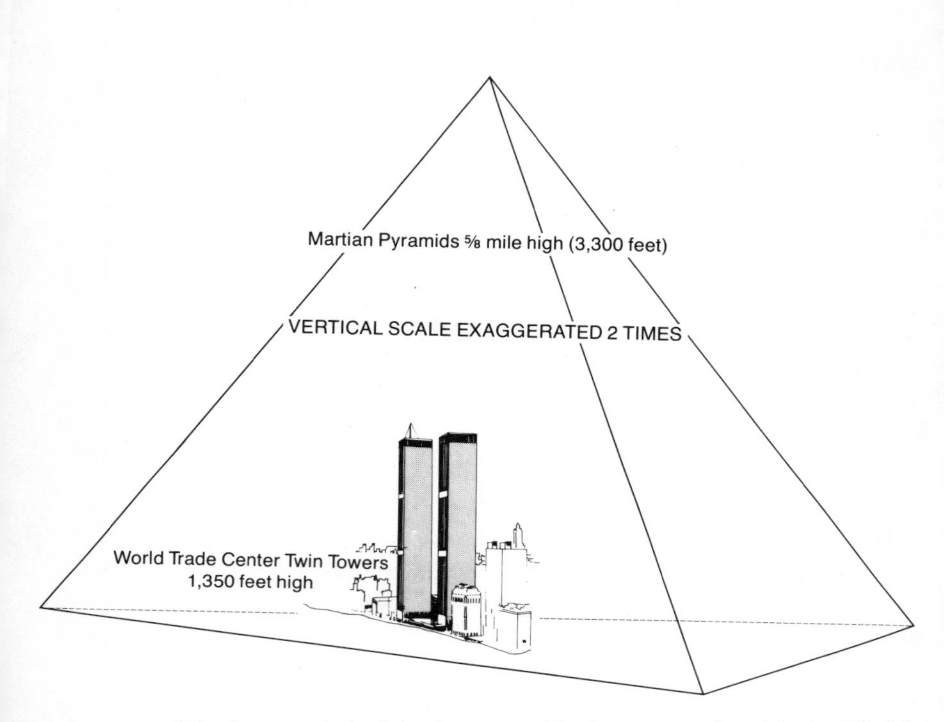

The largest of the Martian pyramids shown to scale with the World Trade Center.

most obvious. A quick examination of the photographs will show that the Martian pyramids, which are distributed on an otherwise flat plain, are oriented in several totally different directions. There are no other features in the area that could have deflected the wind flow between one pyramid and the next. Therefore, if wind faceting were the correct explanation, all of the pyramids would necessarily have been faceted in exactly the same way, and with precisely the same orientation. Since this is manifestly not the case, it seems that this theory is completely untenable.

The second explanation proposed was that the pyramids were formed by glacial sculpturing similar to that which produced alpine "horns" on Earth, the most familiar example being Switzerland's Matterhorn.

There are a number of serious objections. First, there is no evidence for the existence of glaciers anywhere on Mars, and certainly not in this vicinity. Any glaciation massive enough to have produced pyramids would have left other evidence of erosion and deposition. And if glaciers ever existed there, why are they not present now, while Mars is in the depths of extreme ice-age conditions?

Secondly, the glacial processes that carve alpine horns are intimately related to the geological processes of extreme mountainous environments on Earth: crustal folding, upheaval, and compression. There is no indication that such processes take place anywhere on Mars, and certainly not in this region of even, flat desert.

Finally, even if one could swallow the notion of alpine glaciation in the middle of a featureless plain, there is not a single example of such a formation on Earth that comes close to the regular geometry shown in the Martian pyramids. And the irregular alpine horns seem to be isolated features among the mountain peaks, unlike the numerous, closely-spaced pyramids of Elysium.

A third explanation is described as "regolith mantling of erosional remnants of either intersecting resistant dikes, dipping sediments, or other bedrock forms." [2] This means that sand has covered some naturally occurring three-pointed shape, which might have resulted from the intersection of dikes (walls of harder, volcanic rock that formed when molten magma rose up through cracks in the surface material, then solidified). If three such dikes met at one point, and later erosional processes wore away the softer surrounding rock, then three walls would remain standing and might collect enough sand between them to give the impression of flat faces.

One could perhaps imagine the formation of one pyramid by this mechanism, although it seems odd that on Earth, where this is a commonplace geological process, no such shape has ever been found resulting from it. But to go on to the extrapolation of a half-dozen or so of such precise shapes, neatly arrayed in a group, is just asking too much of the laws of probability.

If a group of such configurations *were* to occur, then it would be expected to follow a pattern of interconnecting dikes so that a wall from one pyramid would line up with a wall of an adjacent pyramid. But no such pattern is seen. Indeed, there are cracks that appear to be fault lines in the area, but they all follow an orientation that does not correspond with the orientation of any of the pyramids.

In addition, it is hard to imagine a wind pattern that would be capable of depositing sand against all three sides of an obstacle. Wind patterns generally deposit sand on one side, and scour it away from the other. This is indeed the pattern that is seen in most Martian features that are known to have been caused by aeolian processes, such as wind-eroded craters. There is no apparent way to form flat faces of sand on all sides at once.

The fourth explanation is perhaps the most difficult to accept of all—the "rotation of solidified lava blocks in the underlying molten lava." [3] In other words, somehow exact cubes of solidified lava were formed, which then tilted into a molten lava field just far enough to expose one corner, at which point the surrounding lava solidified to hold the blocks in place.

There are a few obvious problems. Why would the solid lava divide itself into regular cubical chunks with perfectly flat faces? Why wouldn't these chunks be at least partly melted if they were surrounded by molten lava? Why would all of the chunks tilt exactly far enough to expose a symmetrical tetrahedral corner, and then tilt no further? And if these chunks resulted from the breakup of an area of previously formed solidified lava, then what happened to the rest of the chunks?

As in the previous theory, although this scenario is certainly possible, that is, it contradicts no physical laws, it must surely have a very high order of improbability. This whole sequence of events, in which each stage is in itself unlikely, is very difficult to accept.

In sum, no one seems to have been able to come up with a geological explanation that is convincing. Perhaps there was some other mechanism that was unique to Mars, which no one has yet been able to imagine. But there seems to be no reason, given the present lack of any easily acceptable explanation, to exclude from consideration the most obvious conclusion of all—perhaps they were built by intelligent beings.

Looked at in this light, the pyramids would no longer seem to be such an improbable feature. From the point of view of a civilization that wanted to leave some kind of mark, a monument that would last as long as possible to record the existence of the race that built it, the requirements of durability and recognizability as an artifact lead quite logically to the choice of a pyramidal shape. On Earth, four-sided pyramids have been constructed, apparently independently, by a number of ancient civilizations, most notably in Egypt and Mexico. On Mars, three-sided pyramids occur. In principle, there is no reason that other civilizations could not have built pentagonal, hexagonal, or even higher-order pyramids.

The possibilities for speculation are endless, but it is only speculation on limited data. We will not really know the answers until the day when a person can stand at the base of the pyramids of Elysium, gaze up at them, examine their surroundings, study them, and perhaps finally learn the truth about their origins.

Regularity and Repetition

No theory yet proposed accounts convincingly for the formation of these pyramids. But the absence of a geological explanation is not, in itself, evidence for an artificial origin. Are there any criteria by which it is possible to determine from afar whether an object is natural (i.e., results from nonbiological forces such as erosion and volcanism) or artificial (i.e., constructed by intelligent beings)?

The Nobel-Prize winning French biologist Jacques Monod has produced just such a set of objective criteria, designed to enable even a computer to distinguish artifacts from natural objects. There are two criteria, described in Monod's book *Chance and Necessity*: regularity and repetition.

Regularity refers to the fact that, as Monod writes, "natural objects, wrought by the play of physical forces, almost never present geometrically simple and straightforward structures: flat surfaces, for instance, or rectilinear edges, right angles, exact symmetries; whereas artifacts will ordinarily show such features, if only in an approximate or rudimentary manner." [4]

Flat surfaces . . . rectilinear edges . . . exact symmetries—even though these lines were written before the beginning of the Mariner 9 mission, they read almost like a description of the Martian pyramids. And the second criterion is even more appropriate: repetition.

Monod writes: "Of the two criteria, repetition would probably be the more decisive. Materializing a reiterated intent, homologous artifacts meant for the same use reflect, faithfully in the main, the constant purpose of their creator. In that respect *the discovery of numerous specimens of closely related objects would be of high significance*" (emphasis added).[5]

Indeed, it is possible, albeit at some stretch of the imagination, to believe that a single pyramid might have been formed by natural geological processes. But there are at least four distinct pyramids in the photograph, as well as other similar structures. Furthermore, they appear to be paired:

there are two very large pyramids, almost identical in size, standing side by side; nearby, two smaller ones that also seem to be nearly identical in size are also seen side by side. Such a repetition of forms fits perfectly Monod's criterion of "numerous specimens of closely similar objects."

Thus, using the criteria Monod has described, one would have to classify the Martian pyramids as artifacts, objects constructed by intelligent beings.

How reliable are Monod's criteria? Is it possible to fool these "objective" rules, to find an object that satisfies all the requirements, yet which is not an artifact? Monod describes two kinds of exceptions—categories of objects which meet the criteria even though they are not artificial products. The first exception is crystal formations, whose precise geometries and repetitions satisfy the conditions perfectly. Crystals are a unique phenomenon in nature, objects whose large-scale form directly reproduces their molecular structure because the molecules themselves are tightly packed in an orderly array. But crystals require a very precise and stable environment for their formation, and they never reach sizes of more than a few inches. Since the Martian pyramids are two miles across, it is simply not possible that they could be crystal formations. Such a large mass just would not be stable enough for the delicate process of crystal formation.

Monod's second exception concerns living creatures. Almost all living beings possess some degree of symmetry; in the case of most higher animals, it takes the form of bilateral symmetry, which makes a mirror image look almost identical to the creature being reflected. In lower animals, it takes many different forms; for example, the fivefold symmetry of a starfish. Furthermore, repetition is something of a specialty among living organisms: by reproducing themselves, living creatures have produced large populations of nearly identical individuals.

Again, the size of living creatures that we are familiar

with makes this exception unlikely to be relevant to the case of the Martian pyramids: the largest living things on Earth are giant Sequoia trees which reach heights of more than 300 feet. This is still more than ten times smaller than the pyramids, in terms of height, and thousands of times smaller in volume. Nevertheless, it may be just barely possible that some extraordinary, gargantuan life form of an unknown kind might be able to reach such a size. The discovery of such a creature would be hardly less stunning than an artifact of an ancient civilization.

There is one other related exception which Monod cites; it is the construction by living creatures of highly regular forms which are not themselves alive. For example, bees build hives whose endlessly repeated hexagonal cells fulfill all of the criteria. Again, such constructions on Earth never reach sizes of more than a few inches. Nevertheless, there are some African termites which build tall, spindly hills that can reach heights of twenty feet, more than two hundred times the length of an individual termite. By analogy, it is possible to imagine some kind of burrowing Martian creatures which, in a gigantic colony, might create a precisely geometric mound as their home. Such advanced engineering and large-scale cooperation would be, in itself, an astonishing and fascinating discovery.

It seems that we must, therefore, give serious consideration to the possibility that these pyramids might result from some kind of large and highly developed life form. There is even a chance that they may have been the products of intelligent, civilized beings. We cannot prove it, of course; the evidence is still circumstantial. Perhaps, although no exceptions on such a scale exist on Earth, Monod's criteria do not encompass some exotic and unknown process of Martian geology that allows pyramids to occur naturally. Perhaps, by some fluke, this unknown process was repeated often enough to produce a whole field of symmetrical pyramids.

But we *may* have already seen, on the plain of Elysium, the first evidence of extraterrestrial intelligence. Whether the architects of these pyramids were still around or not, further exploration of that area of the Martian surface could not help but expand our horizons and add immeasurably to our knowledge. Even the excavation of relatively recent archeological sites on Earth, to say nothing of those that are far removed in time and in culture, broaden our knowledge of ourselves and our civilization. If Mars turns out to have been inhabited by intelligent beings, the archeology of that alien species might be the most important venture of scientific discovery ever undertaken.

The pyramids of Elysium may, or may not, have been built by beings as intelligent as us. We won't know until we go there for a closer look.

The Final Frontier

We are at a point in history where a proper attention to space, and especially to near space . . . may be absolutely crucial in bringing this world together.

—Margaret Mead

THE EXPLORATION of Mars has so far given no absolute answers, but it has raised questions that we cannot afford not to pursue. The Viking life detection experiments gave strong evidence for the existence of some kind of microbial life; the seasonal growth and decline of dark regions is strongly suggestive of plant life; and the great pyramids in the plain of Elysium have raised the most tantalizing possibility of all, that Mars may once have supported, might even still support, intelligent life. With such extraordinary possibilities waiting to be studied, we certainly must not abandon the search at this point.

Questions about Martian life do not merely tickle our fancy and arouse our natural curiosity and desire to understand everything around us. These questions are important, and the research that we undertake to answer them will have profound consequences for the future of mankind.

Our understanding of biology, of biochemistry, of the

very nature of life itself, is limited. We have only one example, one kind of life to study. Every living thing on Earth is the descendant of a single, primitive cell that came to life in the Earth's primeval slime three billion years ago. To be able to compare details of the construction of terrestrial life with just one Martian bacterium, however primitive, however simple, however unfamiliar, might give us more insight into the very nature of life than any other kind of observation could ever provide.

Just as a fish is unaware of the water because it has never experienced anything else (until it finds itself suspended above it, dangling from a hook), we may be unaware of some of the most fundamentally important characteristics of earthly living things, simply because we have never had any other kind of life to compare them to.

We cannot foresee what new knowledge will emerge from such a comparison, when and if it happens. But we know that every major new discovery about the basic workings of biological systems has resulted in almost immediate practical applications in the improvement of health care and the quality of life.

And yet, with matters of such import waiting to be investigated, at the time of this writing not a single mission of any kind for continuing the exploration of Mars has yet been approved or funded. What a tragedy it would be if no such missions were undertaken! It would be a bit like having Columbus return to the Old World from his voyage of discovery, only to have Europe continue about its business without ever venturing forth again to the Americas. Surely the curious spirit of man cannot be content for long to leave his most fascinating horizon unexplored. Surely, before long, we will return to Mars.

Although no such return is yet scheduled, NASA scientists are hard at work with their plans, ready for the day when funding will be authorized and the long task of preparation can begin. But the preparations for a journey of 400 million miles are extremely involved, so that even if money were appropriated now for the most immediate Mars mission possible, the actual launch would not occur before 1984. Many former Viking scientists are resignedly predicting that a launch date before 1988 is very unlikely, since most of NASA's budget is currently tied up in the shuttle program.

But the planning goes on, with a wide range of possible missions already on the drawing boards. The proposals extend all the way from a sort of "Viking on wheels" that could be ready for a launch in the 1980s, to visions of modifying the entire Martian climate to make the planet suitable for colonization—a plan so audacious and so far in the future that few science fiction writers would have dared to suggest it, and that probably could not be carried out before the twenty-second century.

Marsmobiles

When a mission is approved, it will probably involve a Martian "rover" vehicle of some kind. The proposal currently favored by NASA would consist of two orbiters, two rovers, and several hard-landing penetrators. One proposal had been simply to modify the backup Viking lander, which is already complete and ready to go, by installing wheels or tractor treads to make it mobile. But now the interest has shifted to designing a new vehicle from the ground up, in order to take as much advantage as possible of the information gathered by the Viking mission. This Mars rover might end up looking more like the "Moon-buggy" used by the Apollo astronauts than a Viking lander.

Either way, whether it's a modified Viking or a brand new Marsmobile, the most important feature of this proposed lander is its ability to cross large distances on the Martian surface. This would solve one of the great frustrations of the

Viking mission, which was that the landings had to be made in areas that were deliberately chosen to be as dull as possible, in order to minimize the dangers of landing on rough terrain. The mission scientists had to sit there thinking about the more interesting, biologically promising areas off beyond the horizon, resigned to the fact that the lander was firmly planted in one dull spot, and could never go and peek over the next hill to see what lay beyond. A rover could descend in a smooth area, and then set off toward whichever horizon looked most promising to the controllers back on Earth. The robot craft would be preprogrammed with some basic instructions, like "don't go over a cliff" or "if you get stuck, stay where you are and holler for help." It would also be capable of maneuvering around small obstructions. But its overall itinerary would be planned on Earth, and because of the difficulties of relaying the instructions, it could go no further in one day than the horizon on the previous day's photos. Progress would be slow, but over a six-month mission it might be able to cross 500 miles or more, and to explore a representative variety of Martian topography.[1]

One suggestion is that the rover might land in one of the great equatorial rift valleys, and then climb up its edge to the top, thereby crossing an array of different strata in the rock and building up a detailed picture of the planet's geological history.

Another possible itinerary might involve landing in the smooth plains and then moving up into the mouth of one of the channels. The rover could follow the meandering valley all the way back to its source, crossing a wide variety of terrains along the way while remaining in an environment that, because it was once a riverbed, might be a likely habitat for life: layers of alluvial sediment might, like a terrestrial swamp, be rich in organic matter, and might even now be more moist than most areas of the Martian surface.

It is even possible that some of these river channels have

Artist's conception of a Mars rover mission, which may be launched in the 1980s.

their source in a region where subsurface ice has been melted by volcanic heat. If the rover were to reach such an area by following a riverbed, it might discover a still active hot spring or a zone of slow but constant melting—the kind of "microenvironment" whose possible existence has been suggested as an ideal habitat for life, even for quite advanced forms of life. If such fertile oases do exist in the Martian desert, following the river channels might be the best strategy for finding them.

Another likely itinerary might involve landing near the edge of the polar ice cap in early spring. By moving northwards to follow the dwindling glaciers, the rover would traverse a zone of just-melted ice. This fringe of moist earth has the greatest relative humidity and the greatest amount of available water of any region on Mars today, and is the site of dark markings that may be signs of plant life. Alternatively, the rover could land in one of the equatorial dark areas of active seasonal change, such as Syrtis Major. By roving through these areas during the spring and summer growing season, the cameras of a rover might immediately reveal, once and for all, the reason for the seasonal changes. If windblown dust were the cause, its action would be easily seen. If plants were flourishing in the area, they too would be clearly seen. And if plants *were* seen growing, the rover could go up to them and begin sampling and testing their chemical activity; we would then not only have proved the existence of Martian plant life but would also have an opportunity to explore its nature.

There are many other possible regions for a rover to explore, some with the potential for discovering life forms, and others of primarily geological interest. One might roam around the Hellas basin, the one place on Mars with enough atmospheric pressure for water to exist. Other rovers might explore a volcanic region, or part of the heavily cratered central highlands, or the "etched plains" circling the North pole, where deep gouges were formed in the surface by as-yet-unexplained processes.

With so many potential targets, we would either need a large number of rovers, or some additional way of reconnoitering possible sites. One possibility that is being studied is to accompany the rover with some kind of remote-controlled aircraft.[2]

These aircraft could then be used to scout out areas of possible interest before a final rover itinerary was chosen. In addition, they could themselves be equipped with some instrumentation so that upon landing they could do some experiments of their own on the soil they found.

Two kinds of aircraft that have been proposed are high speed hydrazine-powered propeller aircraft, and lighter-than-air craft, perhaps resembling weather balloons.

The jet planes would have the greatest speed and range, but probably could not land intact on the rough Martian surface. Their one-way kamikaze missions would only be useful for scouting. Balloons, on the other hand, could land to do experiments and then fly off again to other locations. However, balloons would be at the mercy of the winds, and could not choose the most desirable locations for landing and sampling. They would therefore be of little use as scouts for the rovers, and their instrumentation might be wasted on uninteresting landing sites.

The ideal aircraft for the rover mission might therefore be some kind of dirigible. A dirigible would have the advantage of being able to land and take off as often as required, and could go anywhere that seemed promising. Although its speed would be very slow compared to a jet aircraft, its range for a given amount of fuel might be substantially better, since no fuel is used just to keep it aloft. And even after all its fuel was used up, it could continue to drift around the planet at the mercy of the winds, providing additional low-altitude aerial photography.

Whichever of these alternatives is chosen, this low-altitude photography would be an invaluable addition to the mission. It would greatly add to our photographic coverage of the planet, providing a useful intermediate between the total coverage of the planet by low resolution satellite photography, and the tiny fraction of the surface that could be covered by high resolution, ground based photography even from a rover.

Assuming that a rover mission is eventually approved, the biggest decision remaining to be made is the selection of the instruments that it should carry. At the moment, there is a heated debate going on as to whether there should be any life detection experiments included in this mission at all. In fact, the preliminary recommendation of the planning group was *not* to send any specific biology experiments, but to concentrate all efforts on the investigation of geological and chemical processes.

Carl Sagan, director of the Laboratory for Planetary Studies at Cornell, has been the most outspoken critic of this approach. He says that "by the criteria established before the launch, two of the three [biology] experiments came up with positive results . . . There has been no plausible non-biological model put forward which explains the data. Therefore the possibility of life on Mars, detectable by biology instruments, is astonishingly very much open." Therefore, he contends that the idea of returning to Mars without active biology instrumentation "is sheer madness." [3]

Many of the scientists favor using a modified version of the GCMS—the device used on Viking to look for organic molecules—as a simple life detection instrument that would also have many other functions, such as analyzing the chemical composition of the air and the soil. Since everyone agrees that a GCMS should be aboard the rover anyway, all that would be required would be to add a sensor that could be pushed into the Martian soil to sniff for signs of respiration indicated by the appearance or disappearance of certain gases. According to Arden Albee, professor of geology at the California Institute of Technology, "If in the course of roving, we find an 'oasis,' we will have techniques to look at it. We will not have a nutrient experiment, but we will have the capability to detect organic compounds." [4]

But Sagan maintains that this would be inadequate. "The Viking GCMS could not find any organic molecules. Its sensitivity was 1,000 times below the equivalent sensitivity of the two biology experiments which gave positive results. To do the GCMS again is to opt for repeating the experiment that did not work." [5]

Most of the scientists involved in planning this mission are highly skeptical about the prospects for finding Martian biology, and so they are reluctant to invest much in the way of money or resources in the effort to detect it. The danger of this attitude was brought out by James Arnold, a chemistry professor at the University of California who is on the planning team: "The biology stock is down, but we haven't forgotten Mariner 4, when everybody said Mars was a totally dead planet. That was all a bunch of baloney because we had only looked at a narrow strip. On Viking, we sampled only two sites, and there is a very strong analogy between the present situation and the situation after Mariner 4." [6]

Harold Klein, head of the Viking biology team, sums it up thus: "I would find it unacceptable as a citizen-taxpayer to search for a place [likely to support life] and find one, and not have the ability to find out if there is any life there." [7]

If they *do* decide to send a biology instrument, there are a number of possible candidates to choose from. We certainly would not want to use the same experiments that were sent on the Viking mission; although it is possible that they might produce significantly different results in a different location, chances are that any further repetitions of the same tests will do little to clear up the confusion that still surrounds the interpretation of the Viking results. It would certainly be more productive to send a whole new set of experiments that might shed some light on the nature of the reactions that were observed by the Viking tests.

One of the problems in understanding the Viking results stems from the fact that all of the nutrients used in the tests were mixed together, so that there was no way of telling which one or ones of the nutrients were actually involved in the reactions that took place. Sagan points out that "the usual way to conduct science is to vary variables one at a time, and that was not done on Viking. The experiments were so constrained that the results have to be obscure. I'd like to see similar experiments but with a lot more variables and a lot more sample runs . . . If we could do one nutrient at a time to see which ones were used up, that would be very important." [8]

The biology instrument that is currently under development for possible use on the rover would do exactly that. Called the "unified biology instrument," it contains eleven separate test cells, each capable of holding soil samples of from one to ten milliliters (from less than a teaspoonful up to about half an ounce). Each test cell is provided with three containers of nutrient, which can be added one at a time. Thus, the instrument can monitor the reactions to a total of thirty-three different nutrients. The reactions would be monitored by a mass spectrometer that would measure any gases being given off by reactions taking place in the test cells. These gases would be analyzed once a day. In addition, the instrument can vary the temperature at which the incubations take place, and can inject water into any of the cells at any time during the experiment. The whole device takes up about the same amount of space as the Viking biology instrument—about a cubic foot.[9]

The choice of the actual nutrients to be used could be

left until the last minute, so that the latest theories about the meaning of the Viking experiments, and about possible Martian life forms, might be tested. For instance, if biologists had concluded that a particular Viking result was caused by one of the nutrients in the mixture that was used, then that one nutrient could be tested separately to see if the same reaction occurred. If it did, that would prove that it was, indeed, that substance that caused the response, and would make the interpretation of that result much easier.

Of course, mixtures of nutrients could also be used in some of the test cells. One possibility would be to repeat the labeled release test, which gave such strong positive signals, using a nutrient solution in which the stereoisomers had been separated.

Stereoisomers are organic molecules of biological origin that are asymmetrical. They exist in "right-handed" and "left-handed" versions, which are chemically identical but which are mirror images of each other. The significant thing about them is that in most cases living organisms only use one of the versions of each molecule: when organic molecules are made by nonbiological, chemical means, left-handed and right-handed versions are produced equally, but biological systems always select one or the other.

These molecules can be thought of as right- and left-handed gloves. Although they are made in the same way, from the same materials, the right glove simply will not fit on your left hand. In the same way, the right-hand version of an organic molecule, although it is made in the same way and has the same chemical composition, just will not fit into the chemistry of a left-handed biological metabolism.

So let's say we repeat the labeled release experiment with the same mixture of nutrients that were used last time, but separated into two batches—one batch containing only the left-handed molecules, the other batch containing only the right-handed ones. If the reactions produced by these two batches were different from one another, that would prove that biological activity was involved.

Another life detection instrument that has been proposed also makes use of this asymmetry of biological molecules. It is called the Pasteur Probe, after Louis Pasteur who discovered the "handedness" of organic molecules. It differs from the test just described in that, instead of adding nutrients to the soil (which the Martian organisms might not happen to like), it would add a substance that reacts with amino acids in the soil. If we labeled one version of the reagent, either right or left, with a heavier isotope of hydrogen, it would be possible for a GCMS to determine whether the amino acids were right-handed, left-handed, or evenly mixed.[10]

Another possible test for life, developed by Gilbert Levin (who also designed the labeled release test) is an instrument called Diogenes, which would search for signs of ATP (Adenosine triphosphate), a compound that is present in all terrestrial life. The advantage of this test is that it is virtually instantaneous. It is based on the chemistry of the firefly, which produces its light with a mixture of five chemicals, one of which is ATP. So, if the other four chemicals are provided, then any ATP that is added to the mixture will cause it to light up immediately, signaling clearly the presence of this biological compound.[11] Unlike the other tests, this one could monitor the atmosphere for the presence of any airborne organisms. This might be a particularly useful approach if it turns out that the "oasis" theory is correct: although there might not be active life in the immediate vicinity of the rover, it might still detect signs of life that were blown around from a distant patch of thriving life.

Another multipurpose device, similar to the unified biology instrument, has also been proposed. Called the Multivator (for multiple evaluator), it differs from the unified biology instrument mainly in the means used for detecting responses. While that instrument uses a mass spectrometer to monitor gases in the test chambers, the Multivator uses a light and a photocell to detect optical changes. This could allow it to perform experiments involving the rotation of polarized light by stereoisomers, the reaction of certain organic substances with special dyes, the cloudiness of a solution caused by the growth of microorganisms, or fluorescence caused by organic reactions.[12] Such an instrument could conduct a whole range of tests that are entirely different from any that have been done so far, and might therefore provide a considerable amount of new information. It might also be more sensitive to certain kinds of reactions than the mass spectrometer would be. Generally speaking, the more different kinds of biological tests we can conduct on Mars, the greater the chances of success. Furthermore, the results of one set of tests might help to explain others, so that a combination of experiments might yield more information than any of the individual tests could provide alone.

In addition to the chemical tests for life, several other pieces of equipment could be included on the rover to help in the search for living organisms. One of these is a simple microscope using fiber optics which would attach to the rover's camera, allowing it to inspect the fine particles of Martian soil. This would be of great benefit to geologists, since the shape and size of these particles reveal a wealth of information about how they were formed; for example, examining particles from one of the riverbeds could prove that they were produced by water erosion. But the microscope is even more significant for the biologists: one picture might indeed be worth a thousand words, for a single well-taken microphotograph could do more to prove the existence of Martian life than all of the life-seeking experiments that have been done or proposed. A single picture of the structure of a living Martian microbe, or a set of pictures showing it in motion, could end the controversy all at once.

Two other devices that might be included would be helpful in obtaining and preparing samples for the biology instruments. One is a drill, capable of obtaining samples from a depth of more than three feet, as compared to the Viking's samples which only extended to a depth of a few inches. The other is an automated mortar and pestle, capable of grinding up small rocks so that they could be used in the test chambers. While the Viking experiments were limited to fine particles of soil, it may be that most living organisms on Mars are living in the microscopic spaces inside porous rocks. This has been suggested on the basis of studies recently done in Antarctica, where a multitude of different algae, bacteria, and fungi were found living a few millimeters below the surface of the rocks. There, they were protected against the severe cold, but were still close enough to the surface to obtain the sunlight they needed for photosynthesis.[13] In addition, the rock surface would provide perfect protection for a Martian organism from the lethal ultraviolet of the sunlight. Although these rock-dwelling organisms had not yet been discovered at the time of the Viking mission, in retrospect this now seems to have been one of the greatest limitations of that mission: surrounded by rocks that might have been the likeliest place of all for Martian life to be found, the Viking was limited to sifting through the sand. If a mortar and pestle is provided on the rover, the chances of detecting biological responses may be greatly improved.

Even then, we might not find proof of the existence of life on Mars. If life is primarily restricted to microenvironments, we might not be lucky enough to stumble upon them in our first attempt. And unless some truly foolproof detection system is devised (an unlikely prospect) even overwhelmingly positive results from a whole battery of instruments may not be enough to prove the case to everyone's satisfaction. Life is, after all, based on chemistry, and when its remote detection has to rely on an analysis of chemical responses, chemical explanations for those responses may always be possible.

So, unless we are lucky, rovers may not give us the final answer. Somehow, direct laboratory experiments on samples of Martian soil will have to be performed.

The Sample Return

The next step that has been studied is a sample-return mission. This would be considerably more complex and expensive than a rover mission, because it would involve, for the first time ever, a return trip from Mars. That means that the lander module would have to be a much larger and more sophisticated craft than Vikings or rovers. It might resemble the Apollo landing module, part of which blasted off from the moon to rejoin an orbiting satellite, which made the return trip to Earth.

But the Mars sample-return vehicle would have to be much more powerful, since the Martian gravity is more than twice that of the moon. Also, since the sample return would be unmanned, all the complex maneuvering and trajectory calculations involved in the difficult task of making a rendezvous with the orbiter, which were done by the astronauts and by ground-based computer control during the Apollo mission, would have to be done automatically by built-in programs—an extremely difficult task. If anything went wrong with any of these steps—the landing itself, the digging of a sample, the takeoff, the docking with an orbiter, and the return trip—the whole project would have been a total waste; no information would have been gained.

But what if the sample does make it back to Earth? Will that answer, once and for all, the question of whether life exists on Mars? It's not likely. If we were very lucky, we might find unmistakable signs of life in the sample. But, if we found no signs of life at all, that would not necessarily mean anything. Whatever life may have been in the sample originally might simply have died en route, as a result of the intense radiation of outer space, or as a result of being sealed off from light and moisture for the long journey back to Earth. The organisms might simply have starved to death, deprived of some essential nutrient. Or the sample might have been dug in the wrong place; it might have been sterile, even though there was a haven of life just over the hill.

And if there *were* life in the soil sample, despite the randomness of our sampling methods and the rigors of space travel, what might be the consequences? Most likely, no Martian organism could survive under normal Earth conditions. Having evolved on such a different world, our heat, humidity, and heavy oxygen concentration would probably kill them off at once. But we cannot be sure of that. It is possible that, if they escaped into the Earth's environment, they might cause an unprecedented "Andromeda Strain" epidemic among the creatures of this planet. As unlikely as it may be, the result could be so disastrous that the risk is simply not worth taking. It would be folly to take even a small unnecessary risk that could bring about such wholesale destruction. However carefully designed our containment and quarantine procedures were, we could not be sure that the sample return rocket might not crash, or burn up in the atmosphere upon its return to Earth. An otherwise perfectly successful mission might end up raining a plague down upon the world.

If all this seems farfetched, remember the theory put forward by Sir Fred Hoyle: all life may have originated far out in space, and microorganisms may have been raining down ever since onto all the planets. Hoyle believes that this not only accounts for the origin of life on Earth, but that it is also the cause of the Earth's epidemic diseases. New viruses and bacteria from space may be constantly entering the atmosphere; when a new strain arrives to which we do not yet have an immunity, people all over the world suddenly become ill. Hoyle has analyzed the spread of widespread epidemics, some of them dating back to long before the days of rapid transportation, and has concluded that direct contact between individual people cannot explain the rapid spread of disease. There must have been some source that could infect large parts of the globe, or even the entire planet, all at once.[14]

If that is true, then a load of soil from Mars might contain a rich assortment of viable space bacteria, even if it turned out that there was no active, indigenous life there at all. Viruses, even in great abundance, could not have been detected by any of the tests so far conducted on Mars, or by any of the ones proposed for the rover mission. They are not alive; they only "come to life" when they are able to enter a living cell, at which point they take over its functions. They are the hijackers of the biological world. Since we cannot test for their presence, we would have no way of knowing if they were contained in the Martian soil sample.

All things considered, the sample return seems to be the least valuable and most ill-conceived of all the proposals for Martian exploration. This is particularly alarming in view of the fact that some NASA scientists are pressing to go ahead with a sample return right away, instead of doing the rover mission; this would be doubly foolish, since we know so little about Mars at this point.

Astronauts on Mars

Unless the rover mission gets lucky and stumbles on an active site of Martian biological activity, the final answers about life on Mars will probably have to wait for a manned mission. Such a mission would not be the kind of brief commuter trip that we saw in the Apollo moon landings. Where the lunar missions were counted in days, the Mars landing will be counted in months, perhaps in years. A reasonable timetable might be a six-month journey to Mars, three months of exploration, and a six-month return trip. Such a mission would probably involve at least two, and perhaps as many as six, separate spacecraft, carrying anywhere from six to thirty astronauts.

This flight will not begin on the Earth. The ships will be assembled in orbit from parts ferried aloft by the space shuttle so that they will aleady have a head start, and much less fuel will be required for the launch. After the vehicles have been assembled, the astronauts themselves will be ferried up in one final shuttle trip, and will rendezvous in orbit with the vehicle that will be their home for at least a year. Once they are in place, the countdown will begin. The rockets will fire. The voyage will have begun.

The trip to Mars will seem to take forever. So far, no one has ever spent more than ninety-six days in space; and the Soyuz astronauts, who hold that record, at least had plenty of experimental work to keep them occupied. The Mars team will have little real work to do for the whole six months of the journey. They will certainly have to be selected through a series of very careful psychological evaluations, to be able to stand the claustrophobic and boring conditions of the trip. One of the Apollo astronauts compared his flight to spending fourteen days in a pay toilet; the Mars mission will be a bit like spending a year in a Volkswagen camper with two other people.

Although they will suffer from endless boredom, the Mars astronauts will have one advantage over the Soyuz crew because they may not have to be weightless the whole time. By linking two spacecraft together at either end of a long beam, the beam can be spun like a baton to provide a kind of artificial gravity for the crew. This will keep them from getting so out of shape during the trip that they would have difficulty functioning once they arrive on the surface of Mars. Even so, they will have to spend much of their time during the trip exercising to keep up their muscle tone.

Once they arrive, they will go into a "parking" orbit around Mars, and will begin the task of surveying the surface for a likely landing spot. When the site is selected, part of the crew will remain in orbit, while the rest descend in their landing module, or MEM (for Mars Excursion Module, a larger version of the Apollo LEM).

On the surface, they will probably assemble a fold-out base camp that will be a bit like a popup trailer tent. They will deploy a Mars buggy for their explorations. And they will begin a series of experiments to try to unravel some of the mysteries of this strange alien planet.

The team will probably include many specialists. While only one scientist was ever sent to the moon during the Apollo flights—and that was during the last flight—the exploration of Mars is likely to be conducted, from the very beginning, by experts in geology, chemistry, climatology, and biology. Their work will yield more information about Mars than could ever be gained by remote-controlled instrumentation, because humans have judgment, one capacity that no machine is ever likely to have. A person can decide on the spot what course of investigation is most likely to be fruitful, and proceed accordingly. A person can learn from one set of experiments, and apply that knowledge to the decision of what to test next.

The brain is a marvelously intricate and powerful computer, much more capable than any electronic computer yet devised. A human astronaut might catch a glimpse, out of the corner of his eye, of a tiny fossil somewhere off in the distance. The human mind is attuned to patterns, to a fragment of order in the midst of a field of chaos. The astronaut would immediately run over to pick up the fossil, while a rover

Artist's conception of the proposed manned Mars mission, which could take place in the 1990s. Two landing capsules can be seen, along with rovers and other auxiliary equipment.

2

would have proceeded on its way without ever noticing. For a serious exploration of Mars, nothing but human beings will suffice.

Even with a large team of astronauts, Mars is likely to yield up its secrets slowly and grudgingly. Consider how long it has taken us to explore our own planet, and yet, even today, there are regions that may never have been seen by human eyes: the peaks of the Himalayas, the jungle of the Amazon basin (parts of which were mapped last year for the first time, thanks to Earth satellites), the Arctic tundra and the wastes of Antarctica, and the vast expanse of the Sahara desert. We certainly cannot expect to have a thorough knowledge of Mars after one, or even several, quick trips. A real, in-depth knowledge of the planet will take, at best, many decades to achieve.

But even in one trip, our knowledge of Mars will grow enormously. We will learn about its weather, study its geology, and analyze the chemistry of its soil. And we will finally learn the truth about the existence, the nature, and the extent of Martian life.

When their weeks of exploration are over, the landing party will reassemble at the MEM for the trip home. As with the lunar lander, most of the landing module and its equipment will stay on the Martian surface, perhaps to await the next team of explorers, while a small capsule blasts off from the surface to rejoin the orbiting spacecraft. Then, the long trip back to Earth will begin.

But this mission will only have been the beginning. Once humans have set foot on Mars, they will surely return. Teams of astronauts will go back for longer and longer missions of exploration and research. Someday, we may even set up a permanent colony there.

Mars Base One

The prospects for the colonization of Mars depend on two factors: what kind of life we find there, and what kind of resources we find that might be useful for the Earth. If there is any kind of life more advanced than microbes, then it will be crucially important not to do anything that might disturb Mars' ecological balances; the value to science of an entire alien ecosystem is too important to be tampered with. But if it turns out that Mars is rich in some mineral or fuel resources that have become extremely scarce on this planet, the time may come when permanent mining bases will be set up. Even if this turns out to be impractical or uneconomical, chances are that permanent bases will eventually be installed on the Martian surface, perhaps for the purpose of pure scientific research into the Martian environment.

Such a base might bear some resemblance to the research stations currently being operated on Antarctica. It might be subterranean, consisting of a network of caverns carved from the Martian permafrost; this would save on the cost of transporting building materials from the Earth, but it might be somewhat unstable. A safer installation might consist of a series of heavily insulated quonset huts on the surface.

One of the first priorities of the colonists will be to set up an ice-melting station, fed by the virtually limitless supplies of underground ice, to provide the base with a steady supply of water for drinking and washing. They will also set up large greenhouses, stocked with seeds, and perhaps soil, brought from the Earth. Once the greenhouse begins producing the colonists' food, it will be self-perpetuating, producing enough seed so that further supplies from Earth will be unnecessary. Eventually, it should be possible to use some of the Martian soil as well, although it may have to be sterilized first if there is any hazard of contaminating the colonists' food supply with Martian microorganisms.

The greenhouse will serve one essential function other than supplying the colonists' food: it will also supply them with oxygen, as a natural by-product of photosynthesis in the growing plants. Plants need only carbon dioxide and water to grow, and Mars' atmosphere is almost entirely made up of carbon dioxide, and its soil contains large amounts of frozen water. So plant life is the ideal, natural method for transforming the Martian colony into an oxygen-rich haven. Only at the very beginning of the colonists' stay should it be necessary, while the greenhouse is being established, to supply oxygen by means of electrolysis of water, which could be fueled by solar generators.

Such a colony would be a small, precarious outpost in an overwhelmingly harsh environment. Someday, it may become desirable to establish a large, totally self-sufficient colony. Toward that end, NASA recently commissioned a detailed study of how this could be accomplished; [15] the recommendations that were made were so visionary, so futuristic, that they probably could not be seriously contemplated for centuries. But someday the time may come, and according to this NASA study it's all entirely possible. Someday, we may alter the entire climate and atmosphere of Mars, making it warm and pleasant, with an atmosphere rich in oxygen, just as it may have been long in the past.

The methods for accomplishing this would essentially be ways of triggering the shifts in climate that may naturally occur there in some kind of regular cycle. One suggestion is simply to dust the whole planet with thousands of tons of dark pigment, perhaps carbon dust. The increased warming of the surface that would result might be enough to melt the polar ice caps, thickening the atmosphere and triggering another warm epoch. Another plan is to seed the planet with some hardy species of algae that would be capable of surviving the present harsh climate, and letting them grow and reproduce, gradually building up a supply of oxygen in the Martian air. However, either of these methods might require thousands of years to accomplish the desired effect.

In order to do the job more quickly, one possible plan would be to produce a sort of super algae, using genetic engineering techniques that may be developed within the next few decades. This superplant would be designed not only to withstand the Martian climate, but to thrive in it, growing so rapidly that the planet could be transformed in hundreds, rather than thousands, of years.

Of course, even if these far-flung schemes do become possible, we could not even entertain the notion of transforming the entire Martian environment if it were found to support any kind of life that could conceivably be of scientific interest to us. If Mars supports any kind of advanced life, any plan that might endanger it would be unthinkable.

But if Martian life is limited to the lowliest kinds of microbes, or if we find that there are more highly developed organisms, but that they could survive under the changed conditions—maybe even *prefer* them—then it is entirely possible that sometime during the next millennium Mars may become a second home to mankind, a flourishing, verdant, warm, and Earthlike sister world.

The Final Frontier

The exploration of Mars by human beings may well turn out to be the most important, and the most exhilarating, voyage of exploration we have ever undertaken. Our exploration of the moon was the grandest adventure to date, but the moon is a dull, gray, lifeless place. How much more exciting it will be to explore Mars, the planet of surprises. So far, every step in the exploration of this mysterious planet has overturned most of the previously held beliefs about the place, has answered a few questions only to raise dozens more. We can be confident that when the human exploration of Mars begins, the surprises will be dramatic, and they will be many.

Imagine the excitement that will grip the Earth if a scientist-astronaut is able to make a conclusive and unmistakable identification of a Martian microorganism under the microscope. For the first time ever, a human being will then be face to face with an alien species.

Imagine how much more exciting if an exploring party walks over the crest of a hill and sees a valley waving with a lush blanket of plant life. How Earthlike Mars would seem to us then!

And imagine the awe, the gripping exhilaration that would seize us when the first group of explorers makes its way to the great pyramids of Elysium, if they were to find the ruins of a once-flourishing civilization, a culture rivaling the great empires of our own past. How fascinating it might be to dig through the archeological remains of a race as intelligent and sophisticated as the ancient Greeks, but more dissimilar from us than the dolphins. What insights we might gain into the very nature of civilization and of intelligence.

And imagine, for a moment, the ultimate experience. Imagine that the cities of Mars were not mere ruins, that lurking in the entranceway to the pyramid we find—dare I even say it—a living, intelligent Martian.

Mankind would be transformed. We would have passed from the infancy of our global insularity, passed the initiation into a galactic brotherhood. The lessons to be learned cannot even be imagined.

Of course, we do not have any way of knowing if any of this will ever happen. Mars might still, by some miracle of exotic and unknown chemistry, prove to be uninhabited. Even if it is inhabited, life there may never have progressed to an advanced stage. But can we afford not to find out?

Knowing that there is a strong likelihood that there is some life there, knowing that if there is some life then the prospects for advanced life are good, can we simply abandon the exploration of our sister planet and leave our questions unanswered? Surely not. Of all the desires which drive humanity, curiosity is one of the most insatiable. And curiosity in a matter of such cosmic importance, of such potential for opening wide the grandest portals of knowledge, can surely not be denied for long.

Sooner or later, we will explore Mars. Sooner or later, we will unravel its mysteries and solve its enigmas. We will know what kind of fellow creatures share our solar system with us, and that knowledge will give us greater insight into what kind of creatures we may expect to find among the nearby stars. The process has begun and must now inexorably progress to its conclusion.

We have the capacity to conduct this exploration, to answer these questions, within our lifetimes. The manned exploration of Mars could be accomplished before the end of this century. Let us hope that those leaders in whose hands the decision rests will see the importance of this mission and will see to it that no effort is spared and that our exploration will proceed as swiftly and as thoroughly as possible.

When our great-great-great-grandchildren look back at our era, they are not likely to remember the names of our politicians, the wars that we fought, or the many problems that we faced. The one thing about our time that is likely to be remembered, and to reverberate loudly down the corridors of time, is our reaching outward from the planet of our birth. Our time may well be remembered, for centuries and millennia to come, as the age of Mars.

Notes

In the Beginning . . . , pages 1–15.

1. See discussions by Francis Crick, J. R. Platt, L. Orgel, and Carl Sagan in *Communication with Extraterrestrial Intelligence.* Cambridge: MIT Press, 1973.
2. From a speech by Robert Frosch, the newly appointed director of NASA, reported in the *New York Times,* June 24, 1977.
3. *Will the Universe Expand Forever,* by Gott et al., Scientific American, March, 1976.
4. *Cosmology: Man's Place in the Universe,* by Virginia Trimble, American Scientist, January–February, 1977.
5. *The Age of the Elements,* by David Schramm, Scientific American, January, 1974.
6. The object is designated MWC 349. See New Scientist, June 23, 1977, p. 691.
7. *The Companions of Sunlike Stars,* by Helmut Abt, Scientific American, April, 1977.
8. Reported in the *New York Times,* July 23, 1977.
9. These experiments are described in detail in *The Book of Mars,* by Samuel Glasstone, NASA, 1968, and in *The Ascent of Man,* by Jacob Bronowski, Boston: Little, Brown, 1973.

The Spark of Life, pages 16–38.

1. *Intelligent Life in the Universe,* by I. S. Shklovskii and Carl Sagan, Holden-Day, 1966.
2. *Protein Making Machinery Was a Clue to Ancient Organisms,* New Scientist, Dec. 1, 1977.

3. Op. cit., *Intelligent Life in the Universe,* p. 233.
4. Ibid.
5. *The Book of Mars,* by Samuel Glasstone, Washington: NASA, 1968. See also the article under the heading "Life," by Carl Sagan, in the Encyclopedia Britannica 3.
6. Ibid., p. 219.
7. The role of changing conditions as an impetus for evolution is explored in detail in *Until the Sun Dies,* by Robert Jastrow, New York: Norton, 1977.
8. *The Mind of the Dolphin,* by John C. Lilly, Garden City, N.Y.: Doubleday, 1967.
9. *Chance and Necessity,* by Jacques Monod, New York: Knopf, 1971 (Chap. 7). See also note 11.
10. Op. cit., *Mind of the Dolphin.*
11. *Communication With Extraterrestrial Intelligence,* edited by Carl Sagan, Cambridge: MIT Press, 1973.
12. *Paleoneurology and the Evolution of Mind,* by Harry Jerison, Scientific American, January, 1976.
13. *Current Aspects of Exobiology,* Oxford: Pergamon, 1965.
14. The original equation was written as $N^* = R f_p n f_l f_i f_c T$
15. *The Companions of Sunlike Stars,* by Helmut Abt, Scientific American, April, 1977.
16. Op. cit., *The Book of Mars,* pp. 218–219.
17. Op. cit., *Communication with Extraterrestrial Intelligence.*
18. An interesting discussion on this subject appears in Chapter 2 of *The Next Ten Thousand Years,* by Adrian Berry, New York: Saturday Review Press, 1974.
19. *Where Life Begins?* by Chandra Wickramasinghe, New Scientist, April, 1977; Fred Hoyle and C. Wickramasinghe in Nature, January, 1977; and *Lifecloud: The Origin of Life in the Universe,* by Hoyle and Wickramasinghe, London: Dent, 1978.
20. *Organic Matter from Space,* by Brian Mason, Scientific American, March, 1963.

Mars: The Habitable Planet, pages 39–73.

1. *The Solar System,* special issue, Scientific American, September, 1975.
2. *The Atmospheres of Mars and Venus,* by Von R. Eshleman, Scientific American, March, 1969.
3. Op. cit., *The Solar System.*

4. The moon shots covered a distance of about 300,000 miles, while the Vikings traveled about 400,000,000 miles to reach Mars.
5. The Russian probe *Mars 2* crashed in the Fall of 1971 without returning any data. Its sister craft *Mars 3* landed a few weeks later in the middle of a dust storm, and transmitted twenty seconds of blank pictures before going dead. *Mars 6* crashed in 1973, but transmitted some useful information about the atmosphere during its descent.
6. *The Viking Landing Sites: Selection and Certification,* by H. Masursky et al., and *Radar Characteristics of Viking 1 Landing Sites,* by G. Tyler et al., in Science, August, 1976 (special Viking issue).
7. Op. cit., *The Solar System.*
8. *Mars from Mariner 9,* by Bruce C. Murray, Scientific American, January, 1973.
9. The moon also exhibits a similar asymmetry.
10. *Hot Spots on the Earth's Surface,* by Kevin C. Burke and J. Tuzo Wilson, Scientific American, August, 1976.
11. *Phobos and Deimos,* by Joseph Veverka, Scientific American, February, 1977.
12. *Viking First Encounter of Phobos: Preliminary Results,* by Tolson et al., Science, January, 1978.
13. *Intelligent Life in the Universe,* by I. S. Shklovskii and Carl Sagan, Holden-Day, 1966.
14. Op. cit., *Phobos and Deimos.*
15. Science Dimension, January, 1977.
16. *Life on Mars,* by Patrick Moore and Francis Jackson, New York: W. W. Norton, 1965.

The Oceans of Mars, pages 74–99.

1. *Classification and Time of Formation of Martian Channels Based on Viking Data,* by Harold Masursky et al., Journal of Geophysical Research, September 30, 1977. (This issue reprinted as *Scientific Results of the Viking Project,* American Geophysical Union, 1978.)
2. *Isotopic Composition of the Martian Atmosphere,* by Michael McElroy et al., Science, October 1, 1976 (special Viking issue).
3. *The Long Winter Model of Martian Biology: A Speculation,* by Carl Sagan, Icarus, 1971.

4. *Canon of Insolation and the Ice-age Problem,* by Milutin Milankovitch, Washington: National Science Foundation, 1969.
5. Op. cit., *The Long Winter Model.*
6. *The Case of the Missing Sunspots,* by John A. Eddy, Scientific American, May, 1977.
7. Op. cit., *Classification and Time of Formation.*
8. Ibid.
9. *The Geology of Mars,* by Thomas Mutch et al., Princeton, N.J.: Princeton University Press, 1976.
10. Ibid.
11. Ibid., p. 312.
12. *Martian Impact Craters and Emplacement of Ejecta by Surface Flow,* by M. H. Carr et al., Journal of Geophysical Research, September 30, 1977.
13. No extensive statistical survey has yet been done, but all the examples cited by Carr et al. (note 12) fit the borders delineated by Mutch et al. (note 9).
14. Op. cit., *The Geology of Mars,* p. 281.
15. *New York Times,* Iron-Rich Martian Soil, October 8, 1976.
16. *Intelligent Life in the Universe,* by I. S. Shklovskii and Carl Sagan, San Francisco: Holden-Day, 1966.

The Viking Search for Life, pages 100–121.

1. *Where Are We in the Search for Life on Mars?,* Harold Klein interviewed by Rick Reis, Mercury (Journal of the Astronomical Society of the Pacific), March–April, 1977.
2. Ibid., p. 2.
3. *The Viking Biological Investigation: General Aspects,* by Harold Klein, Journal of Geophysical Research, September 30, 1977.
4. Viking press conference, August 7, 1976, as reported in *New York Times,* August 8, 1976.
5. Op. cit., *Viking Biological Investigation.*
6. Op. cit., *Where Are We in the Search?*
7. *The Pyrolitic Release Experiment: Measurement of Carbon Assimilation,* by J. S. Hubbard, Origins of Life, Fall, 1976.
8. *The Viking Carbon Assimilation Experiment: Interim Report,* by Norman Horowitz et al., Science, December 17, 1976 (special Viking issue).
9. *The Carbon Assimilation Experiments,* by Norman Horowitz et al., Journal of Geophysical Research, September 30, 1977.

10. Op. cit., *Where Are We in the Search?*
11. Quoted in *Profile: Carl Sagan,* by Henry S. F. Cooper, Jr., The New Yorker, June 28, 1976.
12. *The Viking Search for Life,* by Norman Horowitz, Scientific American, November, 1977.
13. Letter to the Editor in the Atlantic Monthly, August, 1977, responding to the article *Life on Mars,* by David L. Chandler, June, 1977.
14. Op. cit., *Profile: Carl Sagan,* p. 55.
15. *Gas Changes as Indicators of Biological Activity,* by Vance Oyama et al., Origins of Life, Fall, 1976.
16. Quoted by John Noble Wilford in the *New York Times,* August 1, 1976.
17. *Preliminary Findings of the Viking Gas Exchange Experiments and a Model for Martian Surface Chemistry,* by Vance Oyama et al., Nature, January, 1977.
18. *Viking Labeled Release Biology Experiment: Interim Results,* by Gilbert Levin et al., Science, December, 1976.
19. Quoted by John Noble Wilford in the *New York Times,* August 1, 1976.
20. Ibid.
21. Op. cit., *Where Are We in the Search?*
22. *Viking Labeled Release Experiment,* by Gilbert Levin et al., Journal of Geophysical Research, September 30, 1977.
23. *Until the Sun Dies, by Robert Jastrow,* New York: Norton, 1977.
24. *Life But No Bodies on Mars,* New Scientist, October 14, 1976.

The Parade of the Seasons, pages 122–132.

1. *Martian Albedo Feature Variations with Season: Data of 1971 and 1973,* by C. F. Capen, Jr., Icarus, 1976.
2. *Variable Features on Mars: Preliminary Mariner 9 Television Results,* by Carl Sagan et al., Icarus, 1972.
3. *Spectral Reflectance of Martian Areas During the 1973 Opposition: Photoelectric Filter Photometry 0.33-1.10 m.,* by Thomas McCord et al., Icarus, 1977.
4. Op. cit., *Martian Albedo.*
5. *The Solar Planets,* by V. A. Firsoff, New York: Crane, Russak, 1977.
6. *The Geography of Mars,* by Thomas Mutch et al., Princeton, N.J.: Princeton Univ. Press, 1976.

7. *Life on Mars,* by Patrick Moore and Francis Jackson, New York: Norton, 1965.

Is There Life on Earth?, pages 133–148.

1. *A Search for Life on Earth at Kilometer Resolution,* by S. D. Kilton et al., Icarus, 1966.
2. *A Search for Life on Earth at 100 Meter Resolution,* by Carl Sagan and David Wallace, Icarus, 1971.
3. *Variable Features on Mars. VII. Dark Filamentary Markings,* by J. Veverka, Icarus, 1977.
4. *Mariner 9 Evidence for Wind Erosion in the Equatorial and Mid-latitude Regions of Mars,* by John F. McCauley, Journal of Geophysical Research, 1973.
5. *The Geology of Mars,* by Thomas Mutch et al., Princeton, N.J.: Princeton Univ. Press, 1976.

The Pyramids of Elysium, pages 149–161.

1. *Pyramidal Structures on Mars,* by Mack Gipson, Jr., and Victor Ablordeppey, Icarus, 1974.
2. Ibid., p. 203.
3. Ibid., p. 203.
4. *Chance and Necessity,* by Jacques Monod, New York: Knopf, 1971.
5. Ibid., p. 5.

The Final Frontier, pages 162–190.

1. *JPL Shapes Broad Planetary Program,* by Donald Fink, Aviation Week and Space Technology, August 9, 1976.
2. *RPV's Studies as Mars Observation Platforms,* Aviation Week and Space Technology, August 22, 1977.
3. *Biology Stress for Mars 1984 Debated,* by Craig Covault, Aviation Week and Space Technology, July 25, 1977.
4. Ibid., p. 55.
5. Ibid., p. 58.
6. Ibid., p. 58.
7. Ibid., p. 57.
8. Ibid., p. 57.
9. *Viking Successes Spur Rover Mission,* by Donald Fink, Aviation Week and Space Technology, September 27, 1976.

10. *The Book of Mars,* by Samuel Glasstone, Washington: NASA, 1968.
11. Ibid.
12. Ibid.
13. *Endolithic Blue-Green Algae in the Dry Valleys: Primary Producers in the Antarctic Desert Ecosystem,* by Imre Friedmann and Roseli Ocampo, Science, September 24, 1976.
14. *Epidemics from Space,* by Fred Hoyle and Chandra Wickramasinghe, New Scientist, November 17, 1977.
15. *On the Habitability of Mars: An Approach to Planetary Ecosynthesis,* M. M. Averner and R. D. McElroy, eds., Springfield, Va.: National Technical Information Service, 1976.

Bibliography

Journals

The following are periodicals which either devoted entire issues to reports of Viking Mars mission results, or which carried extensive coverage of the mission during the periods indicated.

Science magazine, special Viking issues: August 27, October 1, December 17, 1976.

Journal of Geophysical Research, special Viking issue: September 30, 1977.

New Scientist, coverage during August–December, 1976.

Science News, coverage during August–December, 1976.

Aviation Week and Space Technology, coverage during June–December, 1976.

Sky and Telescope, coverage from August, 1976 through January, 1977.

Spaceflight (British Interplanetary Society), coverage throughout 1976.

National Geographic, January, 1977.

Maps

Geologic Maps of Mars,
Shaded Relief Maps of Mars,
Topographic Maps of Mars;

This excellent series, based on Mariner 9 imaging, includes 1:25,000,000 maps of the whole planet, 1:5,000,000 quadrangle maps (full coverage in 30 quadrangles), and larger scale maps of selected areas. Several new maps are issued each year in the series, which is now more than half complete.

U.S. Geological Survey, Reston, Va., 1974– .
Moore, Patrick, and Cross, Charles. *Mars* (atlas). New York: Crown, 1973.

Books

ANDERSON, POUL. *Is There Life on Other Worlds?* New York: Crowell-Collier, 1963.

ANTONIADI, E. M. *La Planète Mars.* Paris: Herman et Cie, 1930.

ASIMOV, ISAAC. *Mars: The Red Planet.* New York: Lothram, Lee and Shepard, 1977.

AVERNER, M. M., and MACELROY, R. D., eds. *On the Habitability of Mars: An Approach to Planetary Ecosynthesis.* Washington: NASA (SP-414), 1976.

BERRY, ADRIAN. *The Next Ten Thousand Years.* New York: Saturday Review Press, 1974.

BRADBURY, RAY; CLARKE, ARTHUR C.; MURRAY, BRUCE; SAGAN, CARL; and SULLIVAN, WALTER. *Mars and the Mind of Man.* New York: Harper & Row, 1973.

BRAND, STEWART, ed. *Space Colonies.* New York: Penguin, 1977.

BRONOWSKI, JACOB. *The Ascent of Man.* Boston: Little, Brown, 1973.

CAIDIN, MARTIN. *Destination Mars.* Garden City, L.I.: Doubleday, 1972.

CORLISS, WILLIAM R. *The Viking Mission to Mars.* Washington: NASA (SP-334), 1975.

DOLE, STEPHEN. *Habitable Planets for Man.* Blaisdell, 1964.

DRAKE, FRANK. *Intelligent Life in Space.* New York: Macmillan, 1967.

FIRSOFF, V. A. *The Solar Planets.* New York: Crane, Russak, 1977.

FLAMMARION, CAMILLE. *La Planète Mars et ses Conditions d'Habitabilité.* Paris: Gauthier-Villars, vol. I, 1892, vol. II, 1909.

GAMOW, GEORGE. *The Creation of the Universe.* New York: Viking, 1965.

GLASSTONE, SAMUEL. *The Book of Mars.* Washington: NASA (SP-179), 1968.

HOYLE, FRED, and WICKRAMASINGHE, C., *Lifecloud: The Origin of Life in the Universe.* London: Dent, 1978.

——— *Frontiers of Astronomy.* New York: Harper, 1955.

JACKSON, FRANCIS, and MOORE, PATRICK. *Life in the Universe.* New York: Norton, 1962.

JASTROW, ROBERT. *Until the Sun Dies.* New York: Norton, 1977.

——— *Red Giants and White Dwarfs.* New York: Harper & Row, 1967.

LEVITT, I. M. *A Space Traveler's Guide to Mars.* New York: Holt, 1956.

BIBLIOGRAPHY

LOWELL, PERCIVAL. *Mars*. Boston: Houghton Mifflin, 1895.

——— *Mars and Its Canals*. New York: Macmillan, 1906.

——— *Mars as the Abode of Life*. New York: Macmillan, 1909.

MACVEY, JOHN W. *Whispers from Space*. London: Abelard-Schuman, 1974.

——— *Alone in the Universe?* New York: Macmillan, 1963.

MAMIKUNIAN and BRIGGS, eds. *Current Aspects in Exobiology*. Oxford: Pergamon, 1965.

MICHAUX, C. M., and NEWBURN, R. L., JR., eds. *Mars Scientific Model*. Pasadena: Jet Propulsion Laboratory (JPL-606-1), 1968.

MILANKOVITCH, MILUTIN. *Canon of Insolation and the Ice-age Problem*. Washington: National Science Foundation, 1969.

MONOD, JACQUES. *Chance and Necessity*. New York: Knopf, 1971.

LEY, WILLY, and VON BRAUN, WERNHER. *The Exploration of Mars*. New York: Viking, 1956.

MOORE, PATRICK. *Guide to Mars*. New York: Norton, 1978.

MOORE, PATRICK, and JACKSON, FRANCIS. *Life on Mars*. New York: Norton, 1965.

MUTCH, THOMAS; ARVIDSEN, RAYMOND; HEAD, JAMES, III; JONES, KENNETH; and SAUNDERS, R. STEPHEN. *The Geology of Mars*. Princeton, N.J.: Princeton Univ. Press, 1976.

OPARIN, A. I. *Life: Its Nature, Origin and Development*. New York: Academic Press, 1961.

OVENDEN, MICHAEL. *Life in the Universe*. Garden City, L.I.: Doubleday, 1962.

PICKERING, W. H. *Mars*. Badger, 1921.

PITTENDRIGH, C. S., VISHNIAK, WOLF, and PEARMAN, J. P. T., eds. *Biology and the Exploration of Mars*. Washington: National Academy of Science, 1966.

RICHARDSON, R. S. *Exploring Mars*. New York: McGraw-Hill, 1954.

——— *Mars*. New York: Harcourt, Brace and World, 1964.

RIDPATH, IAN. *Stars and Planets*. London: Hamlyn, 1978.

SAGAN, CARL. *The Dragons of Eden*. New York: Random House, 1977.

SAGAN, CARL, and AGEL, JEROME. *The Cosmic Connection*. Garden City, L.I.: Doubleday, 1973.

SAGAN, CARL, and SHKLOVSKII, IOSEF S. *Intelligent Life in the Universe*. San Francisco: Holden-Day, 1966.

SLIPHER, E. C. *The Photographic Story of Mars*. Cambridge: Sky, 1962.

SMITH, A. U. *Biological Effects of Freezing and Supercooling*. Baltimore: Williams & Wilkins, 1961.

SULLIVAN, WALTER. *We Are Not Alone*. New York: McGraw-Hill, 1966.

DE VAUCOULEURS, G. *Physics of the Planet Mars.* London: Faber and Faber, 1954.

VIKING LANDER IMAGING TEAM. *The Martian Landscape.* Washington: NASA, 1978.

VON BRAUN, WERNHER. *Space Frontier.* New York: Holt, 1971.

WASHBURN, MARK. *Mars At Last!* New York, Putnam, 1977.

YOUNG, R. S. et al. *An Analysis of the Extraterrestrial Life Detection Problems.* Washington: NASA (SP-75), 1965.

Technical Reports

Mariner-Mars 1969, a preliminary report, NASA SP-225, 1969.

Mariner-Mars 1971 Science Data Team, final report, Jet Propulsion Laboratory (JPL-610-243), Pasadena, 1973.

Mariner-Mars 1971 Television Picture Catalog, JPL-33-585, Pasadena, 1974.

Outlook For Space, NASA SP-386, 1976.

Scientific Results of the Viking Project (reprinted from Journal of Geophysical Research, Sept. 30, 1977), American Geophysical Union (Washington), 1978.

Index

INDEX